Environmentalism under Authoritarian Regimes

Since the early 2000s, authoritarianism has risen as an increasingly powerful global phenomenon. This shift has not only social and political implications, but also environmental implications: authoritarian leaders seek to recast the relationship between society and the government in every aspect of public life, including environmental policy. When historians of technology or the environment have investigated the environmental consequences of authoritarian regimes, they have frequently argued that authoritarian regimes have been unable to produce positive environmental results or adjust successfully to global structural change, if they have shown any concern for the environment at all. Put another way, the scholarly consensus holds that authoritarian regimes on both the left and the right generally have demonstrated an anti-environmentalist bias, and when opposed by environmentalist social movements, have succeeded in silencing those voices.

This book explores the theme of environmental politics and authoritarian regimes on both the right and the left. The authors argue that in instances when environmentalist policies offer the possibility of bolstering a country's domestic (nationalist) appeal or its international prestige, authoritarian regimes can endorse and have endorsed environmental protective measures. The collection of essays analyzes environmentalist initiatives pursued by authoritarian regimes, and provides explanations for both the successes and the failures of such regimes, looking at a range of case studies from a number of countries, including Brazil, China, Poland, and Zimbabwe. The volume contributes to the scholarly debate about the social and political preconditions necessary for effective environmental protection.

This book will be of great interest to those studying environmental history and politics, environmental humanities, ecology, and geography.

Stephen Brain is an Associate Professor of History at Mississippi State University. He received his PhD from the University of California, Berkeley in 2007. His first book, *Song of the Forest*, was published in 2011.

Viktor Pál received his PhD at the University of Tampere, Finland. Currently he is an international postdoctoral fellow at the Laboratory for Environmental and Technological History at the Higher Schools of Economics in St Petersburg, Russia. His first book *Technology and the Environment in State-socialist Hungary. An Economic History* was published in 2017.

Routledge Environmental Humanities

Series editors: Paul Warde (University of Cambridge, UK) and Libby Robin (Australian National University)

The *Routledge Environmental Humanities* series is an original and inspiring venture recognizing that today's world agricultural and water crises, ocean pollution and resource depletion, global warming from greenhouse gases, urban sprawl, overpopulation, food insecurity, and environmental justice are all *crises of culture*.

The reality of understanding and finding adaptive solutions to our present and future environmental challenges has shifted the epicenter of environmental studies away from an exclusively scientific and technological framework to one that depends on the human-focused disciplines and ideas of the humanities and allied social sciences.

We thus welcome book proposals from all humanities and social sciences disciplines for an inclusive and interdisciplinary series. We favor manuscripts aimed at an international readership and written in a lively and accessible style. The readership comprises scholars and students from the humanities and social sciences and thoughtful readers concerned about the human dimensions of environmental change.

Environmentalism under Authoritarian Regimes

Myth, Propaganda, Reality

Edited by Stephen Brain and Viktor Pál

Routledge
Taylor & Francis Group
LONDON AND NEW YORK
earthscan
from Routledge

First published 2019 by Routledge

2 Park Square, Milton Park, Abingdon, Oxfordshire OX14 4RN
52 Vanderbilt Avenue, New York, NY 10017

Routledge is an imprint of the Taylor & Francis Group, an informa business

First issued in paperback 2020

British Library Cataloguing-in-Publication Data
A catalogue record for this book is available from the British Library

Library of Congress Cataloging-in-Publication Data
A catalog record has been requested for this book

ISBN: 978-1-138-54328-7 (hbk)
ISBN: 978-0-367-66486-2 (pbk)

Typeset in Goudy
by Wearset Ltd, Boldon, Tyne and Wear

Contents

Illustrations

Figures

Table

Contributors

Tony Andersson is a postdoctoral fellow in the Program in Agrarian Studies at Yale University, where he is completing a book manuscript, Environmentalists with Guns: Conservation, Revolution, and Counterinsurgency in the Petén, Guatemala, 1917–2017. He received his PhD in modern Latin American history at New York University in 2018. More information on his research can be found on his website, www.tonyandersson.com.

Eunice Blavascunas is an Assistant Professor of Anthropology and Environmental Studies at Whitman College. Her recent publications focus on why and how the past troubles nature conservation in Poland and appear in the journals *Political Ecology*, *Historiche Anthropologie*, and *Anthropology of East Europe Review*. She teaches courses in Communism and the Environment, and Political Ecology. Her upcoming book *Of Forests and Time: An Ethnography of the Bialowieza Forest* will be published by Indiana University Press in 2020.

Stephen Brain is an Associate Professor of History at Mississippi State University. He received his PhD from the University of California, Berkeley in 2007 under the direction of Douglas Weiner and Carolyn Merchant. His first book, *Song of the Forest*, was published in 2011. He has published articles in *Environmental History*, *Russian Review*, *Slavic Review*, and *Cold War History*. He is working on the environmental history of Soviet collectivization, and the Soviet effort to build artificial environments in space. Beginning in 2019, he will serve as co-editor of the journal *Environmental History*.

Nathalia Capellini Carvalho de Oliveira is a doctoral candidate in the Department of History in the Université de Versailles—SQY in France with a grant from the CAPES (Brazil). Her research focuses on the environmental history of the Brazilian military dictatorship (1964–1985), especially regarding the Amazon region, infrastructure building, and energy policy. She has also published on water management, environmental conflict and is interested in gendered perspectives for historical analysis.

Jawad Daheur is a researcher at the CNRS (French National Center for Scientific Research) and is currently attached to the Center for Russian, Caucasian and Central European Studies (EHESS, Paris). He obtained his doctoral degree in history from the University of Strasbourg (2016).

Michel Dupuy is an associate researcher at the Institute of Modern and Contemporary History in Paris. He works in two environmental domains, one relating to the environmental history of the GDR and the other in the media treatment of environmental issues on French television.

Carlos Gomez Florentin is a doctoral candidate in the Department of History at Stony Brook University. He holds an MA in Politics from New York University (2009) and a BA in Political Science (Universidad Católica de Asunción, 2006). He is a Professor of History and Politics at Universidad Católica de Asunción and Universidad Nacional de Asunción. Widely published in Paraguay, his books deal with the recent history of Paraguay. He also earned the SSRC-International Dissertation Research Fund IDRF; Tinker; Fulbright; and LAP UCR fellowships.

Ľudovít Hallon is a scientist of the Historical Institute of the Slovak Academy of Sciences in Bratislava. He is the author or co-author of eight monographs as well as dozens of chapters in monographs and scientific studies at home and abroad. Among his most important scientific outputs is the monograph *Slovakia in the Economic Area of Germany 1939–1945* (Bratislava 2015).

Jiří Janáč received his PhD from Eindhoven University of Technology and currently holds a post-doctoral position at the Institute of Contemporary History of the Czech Academy of Sciences in Prague, Czech Republic. His research interests lie primarily in the environmental history of central Europe.

Agata Agnieszka Konczal is a social and cultural anthropologist. Her area of interest lies in environmental anthropology, political ecology, and environmental history. Her doctoral dissertation and recent book examined the role of Polish foresters as they shape and promote their vision of nature. Studying forest districts in Poland she explored the changing relations between environment and people and the impact of these changes on concepts like human, nature, history, knowledge, and resources.

Muchaparara Musemwa, PhD (University of Minnesota-Twin Cities), is an Associate Professor in the Department of History and is currently Head of the School of Social Sciences, at the University of the Witwatersrand, Johannesburg, South Africa.

Viktor Pál received his PhD at the University of Tampere, Finland. Currently he is an international postdoctoral fellow at the Laboratory for

Environmental and Technological History at the Higher Schools of Economics in St Petersburg, Russia. His first book *Technology and the Environment in State-socialist Hungary. An Economic History* was published in 2017.

Leonardo Valenzuela Pérez is a research fellow at the Institute of Advanced Studies of the University of Santiago (IDEA USACH). His work focuses on the political ecology and cultural sustainability of extractive industries and remoteness in Latin America and the polar regions.

Christopher Reed is a PhD candidate at Iowa State University, Ames, Iowa. His areas of study are Chinese Environmental History and Sino-American relations. He received his Master's from the University of Iowa in 2015 and his Master's thesis was entitled "The Hoover Corollary to the Open Door Notes and China's Civil War."

Miroslav Sabol, PhD, is a scientist of the Historical Institute of the Slovak Academy of Sciences in Bratislava. He deals with the history of economics, history of technology, and social history of Slovakia and Czechoslovakia in the twentieth century. He is the author or co-author of three monographs, numerous chapters in monographs, and dozens of scientific studies at home and abroad. Significant is his scientific monograph of the *History of Transport in Slovakia 1938–1948* (Bratislava 2015).

Richard Tucker is an adjunct professor of environmental history at the University of Michigan. His environmental research centered first on the forest history of colonial India, and then on the ecological impact of American hegemony around the tropical world. Recently he has focused on the global environmental history of warfare and military operations.

Acknowledgments

Contributions in this volume are based on papers presented at the "Commanding the Environment or Green Dictatorships? Nature-Culture—Nature-Society Relationships in Authoritarian Regimes" International Seminar held at the Aleksanteri Institute, University of Helsinki, April 27–28, 2017 and organized by Katalin Miklóssy, Emma Hakala and Viktor Pál. The conference was generously supported by the University of Helsinki, Finland. Editors would like to express their very great appreciation to Katalin Miklóssy, the Aleksanteri Institute and the University of Helsinki for hosting that scholarly event.

1 Introduction

Stephen Brain and Viktor Pál

Environmental protection and classical liberalism have a complicated relationship. Economic liberalism (in the form of free-market capitalism) and political liberalism (in the form of participatory democracy) share an emphasis on personal freedom, a respect for the individual, and a trust in the informed judgment of responsible citizens. Over the past five hundred years, these two forms of liberalism have facilitated, and in turn drawn popular support from, dramatic economic growth—therefore increasing tremendously the pressure on the environment. As a result, the most common and effective efforts to enhance environmental quality have focused on enacting laws that decrease the autonomy of individual actors and increase the power of the state to prescribe suitable behaviors. National parks, clean air acts, endangered species laws, and similar initiatives restrict the freedom of citizens and corporations by setting limits on their liberty to exploit the environment as they see fit.[1] The tight link between environmental protection and centralized government control has led critics on the right to liken environmentalists to watermelons—"green on the outside and red on the inside."[2] Anti-environmentalist protest movements in the United States, such as the Sagebrush Rebellion or the Bundy Standoff at the Malheur National Wildlife Refuge, frequently challenge the government's prerogative to set environmental standards, or the right of the government to own land at all.[3] In short, environmentalists and anti-environmentalists alike identify state authority as a central factor, and often *the* primary factor, capable of curbing the freedom of environmental actors. They only disagree about whether such curbs are appropriate.

Scholars studying environmental issues, although rarely adopting a stridently anti-environmental approach, have often appraised environmentalist initiatives in ways broadly consonant with anti-environmentalist protest. One prominent path leading in this direction is what might be called the "dark green" critique, in which scholars contend that environmental protection, unwittingly or otherwise, has served to reinforce existing power disparities.[4] In the dark green critique, national parks have displaced indigenous and other subaltern peoples; environmental limits on economic activity impose costs borne mainly by the relatively powerless; and developing countries (or "backward" segments of a population) are dispossessed for environmental reasons.[5] From this perspective, the state can be a dangerous force working for the benefit of the powerful, always holding the potential to crush dissidents and subalterns, and observers must remain vigilant so that state authority remains within proper bounds. One of the most influential books in this vein is James C. Scott's *Seeing Like a State: How Certain Schemes to Improve the Human Condition Have Failed*.[6] Scott contends that bureaucratic regimes, generally speaking, aim to organize their societies according to the technocratic principles of "high modernism," failing to take local knowledge, or *mētis*, into account, thereby dooming their projects to failure and their societies to oppression. To be fair to Scott, it is incorrect to label him as a classical laissez-faire liberal. He noted in an online symposium dedicated to *Seeing Like a State* that, in his view, "large scale capitalism is just as much an agency of homogenization, uniformity, grids, and heroic simplification as the state, with the difference that, for capitalists, simplification must pay."[7] Despite this caveat, Scott's main contentions are clear: that centralized planning tends toward a dangerous form of simplification, and that the multifarious crowd can devise better solutions than trained experts. Commenters have critiqued Scott for his propensity to interpret historical episodes tendentiously; one such critic, Russell Hardin, noted that the central problem in one of Scott's key case studies, Soviet agricultural collectivization, was "a failure of incentives, not a failure to rely on local knowledge."[8] Despite these objections, Scott's deeply liberal celebration of local knowledge and individual autonomy, resonant with prevailing Western attitudes toward personal liberty, helped fashion a widely encountered trope in recent environmental history and environmental geography that makes egalitarian subjectivity, rather than the relationship between humans and the natural world, the primary consideration.[9]

The tendency among Western scholars to associate praiseworthy environmental policy with liberalism and individualism is also discernible in the extant environmental histories about authoritarian regimes. Although the environmental policies of smaller authoritarian countries have basically gone unstudied, the available environmental analyses of powerful authoritarian countries such as the USSR, communist China, and Nazi Germany most often focus on negative aspects while casting positive developments in the darkest possible light or omitting them entirely, frequently positing the lack of democratic input into the decision-making process as a key factor.[10] Perhaps this trend is most notable with respect to the country with the best developed environmental historiography, the Soviet Union. The earliest studies dedicated to Soviet environmental policy adopted a measured tone and identified structural problems that resulted in environmental problems, but in the wake of the fall of the Soviet Union, a much more strident approach emerged, centered on Murray Feshbach and Alfred Friendly's *Ecocide in the USSR*.[11] Feshbach and Friendly contended not only that environmental degradation in the Soviet Union exceeded that in other countries, but that the damage done was intentional, hence the term "ecocide," echoing the word homicide.[12] The "ecocide thesis" came to dominate the scholarly discussion about Soviet environmental quality such that it entered general textbooks about Soviet history:

> The rush to modernity ... meant that attention was paid almost exclusively to output and productivity and almost no notice was taken of the impact of rapid industrialization on the natural environment. This insensitivity to the limits of nature was characteristic of capitalist industrialization as well, but in the Soviet Union general ecological ignorance was compounded by the bravado of the Communists, who looked upon nature simply as an obstacle to be overcome on the road to progress.[13]

Although made specifically in reference to the Soviet Union, such a conclusion reflects the tenor of most Western scholarly investigations of authoritarian environmental politics: rarely providing specific comparisons with environmental conditions in contemporaneous liberal societies, but nevertheless confident in asserting that illiberal regimes are quantitatively more damaging and philosophically hostile to the environment.

In contrast, this volume takes as its points of departure that illiberal societies have developed environmentalist policies of their own, that environmentalism is a protean ideology, and that the sets of structures and priorities prevailing in the West represent only some of many possibilities.[14] The essays presented here present two alternative explanations for this phenomenon. The first of these observes that because environmentalism, despite its reputation as a left-leaning political movement, is motivated by a powerful expert-driven, conservative impulse to preserve the world as it is now or restore it to an earlier state, illiberal governments find certain environmentalist policies attractive. Alternatively, the second rationale rejects the sharp distinction often encountered in scholarly analyses between authoritarian governments and the societies from which they spring, and draws examples from illiberal governments implementing ideas informed by popular ideas about nature that might not immediately seem environmentalist to a Western reader. This volume seeks to demonstrate that environmentalism is not a political movement associated with one end of the ideological spectrum, and that authoritarian forms of environmentalism deserve an impartial assessment.

In the opening essay, Jawad Daheur examines the case of a nineteenth-century authoritarian state, that of Hohenzollern Prussia, and finds that the elite's self-assigned role of defending the public good compelled them to abandon their laissez-faire approach to the economy and take control of Prussian reforestation. State officials believed that, left to their own devices, neither Prussian landowners nor yeoman farmers were capable of maintaining the necessary long-term perspective, and thus embarked on a program of mandatory afforestation and land expropriation in places where land shortages militated against forest growth. In some cases, this inflamed ethnic tensions and gave rise to separatist sentiment, but the growing Prussian faith in technocratic paternalism gave rise to a burgeoning forest bureaucracy capable of stopping and then reversing the trend toward deforestation in the early nineteenth century.

In Chapter 3, Tony Andersson analyzes the complex environmental history of the Petén region of northern Guatemala. This heavily forested green corridor has been used for various purposes, including conservation, during Guatemala's decades-long civil war. Andersson focuses on the very difficult problem of conservation in areas such as the Petén, which are dominated by political instability and violence. When conservationist programs, developed in entirely different social and economic circumstances, were implemented in the Petén, they were appropriated by the Guatemalan military command and the Petén's local elite, and then used against the rural peasant population. The misuse of the conservation agenda included the personal sale of officially protected natural resources, including valuable timber and locally produced illegal drugs. Andersson's Guatemalan study resonates well with the economic and environmental history of the Chilean "Chicago Boys" discussed in Chapter 5. That is, when economic and environmental policy measures are borrowed from a different socioeconomic setting and implemented with partial or complete disregard to local conditions, the results are unpredictable if not plainly disastrous. Andersson's study suggests that the human suffering and environmental misuse in the Petén could and should have been mitigated by foreign actors, particularly by the CIA, the government of the United States, and the Western conservation NGOs working with the Petén.

In the next chapter, Nathalia Capellini Carvalho de Oliveira and Carlos Gomez Florentin analyze the complicated environmental history of the Paraguayan and Brazilian joint megaproject, the Itaipu Dam. Developmentalism and symbolic prometheanism have been intertwined with the environmental history of the twentieth century, and megaprojects seemed to attract particular attention from authoritarian regimes. Autocrats used large-scale infrastructural undertakings as symbols of power, prosperity and modernity. Several of these projects, such as the White Sea Canal and the extensive system of Soviet era irrigation canals in Central Asia, were realized with the cost of extensive environmental degradation and human suffering. What Capellini Carvalho de Oliveira and Gomez Florentin illustrate in their chapter is a development story that begins like other megalomaniac infrastructural projects, but over time is derailed by a complex dialog of activist and local communities, international actors, and change within the state bureaucracy. The complex politicization of the Itaipu Dam again suggests that the relationships between the environment and society, even under a military government, are very complex.

In Chapter 5, Leonardo Valenzuela Pérez analyzes a perhaps even more nuanced economic-environmental discourse in Chile under Pinochet's military junta. Despite the stark ideological and economic policy differences between right-wing and left-wing authoritarian regimes, the Chilean government's approach to tackling air pollution problems in Quintero Bay show several similarities with the communist regimes analyzed in Chapters 7 through 10, especially the environmental history of East German air pollution. Despite similarities, the Chilean military junta's economic and environmental policies were built on neoliberal principles driven by the "Chicago Boys," a group of professionals educated at the University of Chicago who returned to their native Chile to serve in high-ranking administrative positions. The Chicago Boys advocated economic deregulation and free-market capitalism, values often associated with liberal democracies. In the case with the Chilean military junta, deregulation and non-intervention in environmental issues became the preferred environmental policy method, shifting the responsibility for environmental protection to private corporations. As was true for East Germany, insufficient funds led to disappointing results, but in the Chilean case, the political will at the elite level was missing.

In Chapter 6, Eunice Blavascunas and Agata Agnieszka Konczal analyze an intriguing environmentalist conflict in post-Communist Poland over the fate of the Białowieża forest. In Poland's highly politicized environmental setting, political trends, religious beliefs and socioeconomic claims produced the confusing cacophony of voices, which Blavascunas and Konczal make legible by grouping them into similar functional roles, regardless of the familiar left-versus-right and green-versus-growth divides. Actors opposing the government's environmentalist agenda were branded as enemies of the state and nation, and as representatives of hostile foreign interests. In contrast, conservationists identified themselves with the global ecological movement transcending national boundaries, and denying the nationalist logic exercised by the Polish government and state foresters. In this heated discourse, the European Union figures as a hostile, alien force, attacking Poland and the Polish people—at least according to some of the participating actors. Most interesting in this environmental controversy is that all sides are convinced of the sole legitimacy and correctness of their environmentalist agendas, although in this instance the positions they take have little to do with the actual conservation of the Białowieża Forest. Blavascunas and Konczal therefore underscore the protean quality of environmentalism and demonstrate that the relationship of environmentalism and authoritarianism is more complicated than is generally understood.

In Chapter 7, the volume's first essay dedicated to a left-wing authoritarian regime, Jiří Janáč explores a complex environmental and technological history in a central European context by analyzing Czechoslovak water and soil improvement efforts under communism. Janáč points out that the Czechoslovak state agencies did not aim for the domination of land and water, but rather what they called a "harmonious" development of the "human environment" which, from the perspective of state agencies, should serve both society and the environment. In Janáč's analysis, the Czechoslovak state was far from James Scott's monolithic interpretation. Rather, state agencies competed and cooperated with each other, as well as with other, independent actors in a dynamically expanding space for discussion. According to this interpretation, state agencies were influenced by, and relied on, the empirical analysis of scientific circles. In other words, rather than imposing its own agenda, the Czechoslovak state delegated technological–environmental questions to the country's scientific intelligentsia that had, in turn, rather diverse ideas about how to develop Czechoslovakia while protecting its natural environment. The result was a bitter turf war and active scientific debates between various groups, with developers and conservationists on both extremes of the scale present in state agencies and society at large.

In Chapter 8, L'udovít Hallon and Miroslav Sabol analyze environmentalism in Slovakia under communist rule, and like Janáč, they underline the environmentalist role of Czechoslovak state agencies, especially with respect to reforestation of Slovakia. Pro-forest state policies were certainly not unique to Czechoslovakia; extensive forest belts were created under Stalin in the Soviet Union, and Reed's essay provides evidence about afforestation efforts in communist China. In environmental questions aside from afforestation, however, the role of Slovak state agencies was ambivalent or explicitly negative toward the natural environment, including the use of herbicides and pesticides in agriculture. Unlike Janáč, who stresses the role of non-state actors and lobby groups, Hallon and Sabol place the constantly changing responses of the state bureaucracy into the center of their analysis. The rapidly changing environmental agendas of the Czechoslovak and Slovak governments reflected changing public opinion and led to adjustments in environmental policy throughout the communist period. Hallon and Sabol suggest that eventually an environmentalist shift within the state bureaucracy took place at a relatively rapidly pace. This swing, most of all, was triggered by the environmental impact of a Stalinist aluminum plant in central Slovakia and an oil refinery combine near the Slovak capital of Bratislava. Ultimately they portray a state apparatus actively seeking to maximize economic growth while protecting environmental quality, in a time of increasing popular expectations.

In Chapter 9, Michel Dupuy explores air pollution problems in East Germany. The build-up of pollution and state environmental control attempts in East Germany resembled patterns from the Czechoslovakian case analyzed in Chapters 7 and 8. Soon after the onset of the Cold War, the East German public began to express concern about the country's deteriorating environmental conditions, and these were in turn acknowledged by the country's political elite. According to Dupuy's analysis, East German environmental problems were very difficult to tackle because much of the pollution resulted from the country's heavy industrial development program and its rigid industrial structure. Nevertheless, Dupuy's analysis suggests that a relative lack of success did not necessarily mean inaction or disregard by the political elite of the country. As the economies of eastern Europe began to stagnate in the 1970s, communist governments found it ever harder to acquire and install up-to-date scrubbers and other cleaning technologies. Shrinking income levels and an escalating economic crisis allowed East German decision makers to focus on a handful of environmental hot spots, and therefore a whole range of environmental issues were left without funding and hence without solutions, although not without attention.

Chapter 10 presents Christopher Reed's challenge to Judith Shapiro's "War against Nature" thesis, and suggests a more complex relationship between Chinese society and the natural environment under Mao Zedong than has previously been posited. He provides evidence from Chinese state-run media that Maoist government state agencies encouraged a protective attitude toward the natural environment, and toward forests in particular, whose health was taken seriously by government officials and forestry workers alike. Citing statistics showing that the rate of deforestation in China did not differ significantly from that of Europe and the United States at the same time, Reed calls into question the historiographical commonplace that Maoism and environmentalism were incompatible. The contrast between the existing characterizations of Communist China's environmental policies and Reed's findings leads him to allege that the historical record has been misinterpreted to fit the agendas of individual researchers.

The volume's last chapter by Muchaparara Musemwa is dedicated to the controversial environmental legacy of Zimbabwe's recently toppled dictator, Robert Mugabe. Environmental issues have been especially central to Zimbabwe's African majority population because of the economic and environmental inequalities that white settler colonialism produced. Immediately after seizing power Mugabe sensed the social importance of environmental issues and used them effectively to gain popular support. However, when around year 2000 Zimbabwe's economy began to decline, environmental concerns were pushed to the background and the country's resources began to be extracted at an alarming rate. According to Musemwa, the rule of Mugabe was a textbook case of mixed environmental values that included both state-sponsored conservation and the extraction of natural resources. Such mixed environmental values and policies were intertwined with Zimbabwe's economic, social, and political changes.

In *A Little Corner of Freedom*, Douglas Weiner's influential book about Soviet nature preserves, Weiner asserts that nature protection "does not exist as a disembodied eternal ideal," and is only comprehensible as a "cultural institution functioning in very specific contexts of space, time and political economy."[15] If this is so, it should not be surprising that authoritarian regimes, despite their illiberalism, have frequently made space for environmentalist initiatives—although the specific contexts of their development shaped the results. In the essays that follow, some commonalities that unite these initiatives will emerge. First of all, forested lands were areas of special concern for many illiberal regimes. In China, Zimbabwe, Guatemala, and Prussia, the government endowed forests with special meaning; explicitly the reasons given for forest protection were environmental, although deeper cultural reasons likely came into play. Second, the authoritarian regimes discussed here were not motivated by hostility to the natural world, nor did they seek to control it in a way vastly different from land-use regimes in liberal countries. Planners in authoritarian countries recognized environmental challenges and accepted their validity, but often found state resources too strained to properly finance environmental protection; better pollution abatement measures were available but too expensive. Third, the environmental protection regimes installed by authoritarian states met with an appreciable amount of success, although often this success did not necessarily equate to increases in liberty or equality for their citizens. Authoritarian regimes, invested with great latitude in pursuing their objectives, have been able to produce effective results when allocating sufficient resources. In communist and non-communist countries alike, environmental protection measures enjoying the full backing of the state successfully re-established forests and set aside protected areas. Again, the primary impediment to environmental policies was not political will or scientific expertise, but economic resources. Richer countries with higher rates of resource consumption were able to ameliorate or outsource their environmental problems more effectively. This raises an interesting question: is increased resource consumption the solution to world environmental problems?

The authoritarian regimes discussed in the following essays vary significantly in geographical location and chronological setting, but they share an engagement with the non-human world that has not previously been discussed in scholarly literature. It is the hope of the editors that the questions raised by the volume will inspire further research.

Notes

1 Even in unusual instances of environmental protection, such as when Ecuador amended its constitution to recognize the Vilcabamba River as a legal entity, or when New Zealand followed suit four years later, the active mechanism is state apparatus, police power, and threat of legal action. See Maria Valeria Berros, "Defending Rivers: Vilcabamba in the South of Ecuador," *Rachel Carson Perspectives* no. 6 (2017): 37–44; Garth Harmsworth, Shaun Awatere and Mahuru Robb, "Indigenous Maori Values and Perspectives to Inform Freshwater Management in Aotearoa-New Zealand," *Ecology and Society* 21, no. 4 (December 2016).

2 James Delingpole, *Watermelons: How Environmentalists are Killing the Planet, Destroying the Economy and Stealing Your Children's Future* (London: Biteback Publishing, 2012).

3 William L. Graf, *Wilderness Preservation and the Sagebrush Rebellions* (New York: Rowman & Littlefield, 1990); James Morton Turner, "'The Specter of Environmentalism': Wilderness, Environmental Politics, and the Evolution of the New Right," *The Journal of American History* 96, no. 1 (June 2009): 123–148.

4 See Stephen Brain, "Dark Green: The Gloomy Side of Environmentalism," *Historical Studies in the Natural Sciences* 45, no. 2 (Winter 2014): 340–347.

5 A comprehensive list of works that draw attention to environmentalism's capacity for advancing the interests of the powerful while feigning selflessness would be too extensive to provide here, but the trend was initiated by essays and books such as William Cronon, "The Trouble with Wilderness," in *Uncommon Ground: Rethinking the Human Place in Nature* (New York: W. W. Norton & Co., 1995), 69–90; Karl Jacoby, *Crimes against Nature* (Berkeley, CA: University of California Press, 2003); Douglas Weiner, "A Death-Defying Attempt to Articulate a Coherent Definition of Environmental History," *Environmental History* 10 (2005): 404–421; Shepard Krech III, *The Ecological Indian: Myth and History* (New York: W. W. Norton & Co., 2000); and Roderick P. Neumann, *Imposing Wilderness: Struggles over Livelihood and Nature Preservation in Africa* (Berkeley, CA: University of California Press, 1998).

6 James C. Scott, *Seeing Like a State* (New Haven, CT: Yale University Press, 1998).

7 James C. Scott, "The Trouble with the View from Above," *Cato Unbound: A Journal of Debate*. www.cato-unbound.org/2010/09/08/james-c-scott/trouble-view-above. Accessed: March 15, 2018.

8 Russell Hardin, "Books in Review: James C. Scott's *Seeing Like a State*," *The Good Society* 10, no. 2 (2001): 39.

9 This trend is especially noteworthy in the field of "political ecology," whose exponents contend that environmental problems are at root political failures to allocate resources fairly. See Paul Robbins, *Political Ecology: A Critical Introduction* (Malden, MA: Blackwell Publishing, 2012); Joan Martinez-Allier, *The Environmentalism of the Poor: A Study of Ecological Conflicts and Valuation* (New York: Oxford University Press, 2005); Richard Peet and Michael Watts (eds.), *Liberation Ecologies: Environment, Development, Social Movements* (Abingdon: Taylor and Francis, 1996).

10 The literature related to authoritarianism as a general political phenomenon is far too vast to survey here, but it should be noted that most environmental histories of authoritarian regimes have adopted an approach inspired by the earliest theoretical works about authoritarian regimes, which conceptualized these governments as operating without popular support and employing terror to maintain power. For general works, see Hannah Arendt, *The Origins of Totalitarianism* (New York: Harcourt, Brace and Co., 1951); Carl Friedrich and Zbigniew Brzezinski, *Totalitarian Dictatorship and Autocracy* (New York: Praeger Press, 1956); Alex Inkeles and Raymond Bauer, *The Soviet Citizen: Daily Life in a Totalitarian Society* (Cambridge, MA: Harvard University Press, 1959); Juan Linz, "Authoritarianism," in *The Oxford Companion to Politics of the World* (New York: Oxford University Press, 1993). For works related to the environmental history of authoritarian countries, see Franz-Josef Brüggemeier, Mark Cioc, and Thomas Zeller (eds.), *How Green Were the Nazis?* (Athens, OH: Ohio University Press, 2005); Judith Shapiro, *Mao's War against Nature* (New York: Cambridge University Press, 2001); Paul R. Josephson, *The Conquest of the Russian Arctic* (Cambridge, MA: Harvard University Press, 2014); John McCannon, "To Storm the Arctic: Soviet Polar Expeditions and Public Visions of Nature in the USSR, 1932–1939," *Ecumene* 2, no. 1 (January, 1995): 15–31; Bernd Stevens Richter, "Nature Mastered by Man: Ideology and Water in the Soviet Union," *Environment and History* 3, no. 1 (1997): 69–96; Thomas Lekan, *Imagining the Nation in*

Nature: Landscape Preservation and German Identity 1886–1945 (Cambridge, MA: Harvard University Press, 2003); Frank Uekötter, *The Green and the Brown: A History of Conservation in Nazi Germany* (New York: Cambridge University Press, 2006).

11 The earliest works on Soviet environmental history include Marshall I. Goldman, *The Spoils of Progress: Environmental Pollution in the Soviet Union* (Cambridge, MA: The MIT Press, 1972); Philip R. Pryde, *Conservation in the Soviet Union* (New York: Cambridge University Press, 1972); Joan DeBardeleben, *The Environment and Marxism-Leninism: The Soviet and East German Experience* (Boulder, CO: Westview Press, 1985); Philip R. Pryde, *Environmental Management in the Soviet Union* (Cambridge: Cambridge University Press, 1991).

12 Books advancing the "ecocide" thesis include Gary E. McCuen and Ronald P. Swanson, *Toxic Nightmare: Ecocide in the USSR & Eastern Europe* (Hudson, WI: G. E. McCuen Publications, 1993); Murray Feshbach and Alfred Friendly, Jr., *Ecocide in the USSR* (New York: HarperCollins, 1988); Feshbach's *Ecological Disaster* (Washington, DC: Brookings Institute Press, 1995); D. J. Peterson, *Troubled Lands: The Legacy of Soviet Environmental Destruction* (Boulder, CO: Westview Press, 1993); Boris Komarov [Ze'ev Vol'fson], *The Destruction of Nature in the Soviet Union* (White Plains: M. E. Sharpe, 1980).

13 Ronald Suny, *The Soviet Experiment* (New York: Oxford University Press, 1998), 238–239.

14 For the purposes of this chapter, the term "environmentalism" will be used to refer to the political and philosophical position that the health of the natural world should be an item of social concern, and that humans should forgo certain practices, especially forms of economic exploitation, to protect the functioning of the natural world as its proponents understand it.

15 Douglas Weiner, *A Little Corner of Freedom* (Berkeley, CA: University of California Press, 1999), 448.

Part I

On the right

2 Reforestation of wastelands and "environmental statehood" in nineteenth-century Prussia

Jawad Daheur

Introduction

With the joint authority, feudal and bureaucratic, on which its monarchy was based, Prussia never pretended to be, or tried to become, a democratic state. It was marked by its hierarchical structure, a high degree of state interference, and the state's claim to work for the welfare of its citizens.[1] Regarding the environment, however, Prussian attitudes have long been characterized by economic liberalism and a low level of state involvement. During the Napoleonic wars, Prussian statesmen emerged as spokespersons of laissez-faire and the reform movement, which started in 1806, earned the civil service a liberal reputation. In 1811, private forest ownership was made entirely free from any state interference, a decision which was maintained until the 1870s, when supervision measures were introduced by law.[2] Meanwhile, this liberal policy turned out to be the cause of serious environmental problems such as erosion, sand drifts and dust storms in connection with the decline in forest cover in many parts of the Kingdom. In 1868, a survey of Prussian agriculture by the agronomist and statistician August Meitzen (1822–1910) pointed out the problem that barren waste and desolate heathlands represented a serious danger to the agricultural productivity of the country.[3] With the ideology of private ownership, long-term forest conservation posed significant questions. These included the issue of who would protect forests if property was privately held and rapidly changed hands, and what responsibility the state had to manage the nation's resources. This chapter tells the story of how environmental change, and the evolving conception of the state during the nineteenth century, prompted Prussian officials to promote the restocking of woodlands that had been depleted through deforestation during the previous decades and centuries.

While during the eighteenth century there was a growing predilection for converting forests into arable land, in the Prussia of the early nineteenth century a broad consensus emerged in favor of reforestation.[4] Tree planting, along with the broader scope of forestry practices, provided the raw material that the national economy needed and was presented as a major factor in promoting economic development for the benefit of both local communities and the nation. At that time, Prussia (alongside Saxony) became the pioneer in the science of managing forests and tree plantations, and during the following period, the academy of Eberswalde, situated near to Berlin, cast its influence all over Europe. Prussian forestry of that time was characterized by a preference for the high forest (*Hochwald*) and a "sustained use" approach which allowed the cutting of only as much timber each year as was planted.[5] Initiatives in reforestation, however, were triggered by a number of factors that do not relate directly to forestry. During the nineteenth century, agronomists, botanists, foresters, and other scientists started to explore the relationships between forest cover, climate, healthy watersheds and agriculture productivity. They claimed that woodlands performed a number of critical services for agriculture. In particular, tree planting would protect against avalanches and torrents, halt desiccating winds, cool and dampen the climate, and improve soil quality.[6] The reforestation movement had three main objectives: timber production, soil erosion control, and job creation in poor areas. Yet, despite all the benefits attributed to forests, the tendency toward clearing woodlands was reversed only in the second half of the nineteenth century and this reversal would not have been possible without a massive commitment of the state.

In many Prussian provinces, endemic poverty, bare soils, and harsh climate conditions rendered the attempts at reforestation difficult and expensive. The involvement of the Prussian administration in the reforestation process after 1850, either by planting its own holdings, by purchasing wastelands for this aim, or by sponsoring reforestation by private landowners, shows that the history of Prussian forests and wastelands should be read in terms of a growing interventionism of the state. In his book *The Political Ecology of the State: The Basis and the Evolution of Environmental Statehood*, the geographer Antonio Ioris not only sees the contemporary state as the main force behind environmental change, but also seeks to understand how the reactions to environmental problems have played a crucial role in the modernization of the state apparatus. He argues that the political philosophies of Hobbes, Kant, and Hegel hold the key to understanding what he calls "environmental statehood," which is "the combination of discursive, ideological, and material efforts by the state to deal with socio ecological problems."[7] For Ioris, the environmental state began as a Hobbesian Leviathan, protecting the environment from a civil society that would surely destroy it if left alone. As the driving force behind the unification of Germany in 1871 and the leading state of the German Empire (accounting for two-thirds of its territory and population), the Kingdom of Prussia offers an interesting case study of the linkages between the reforestation movement and environmental statehood. A fully organized state forest administration existed from the 1820s, when directive, inspection, and executive services were established.[8] But it took a long time for the Prussian state to take charge of the reforestation of wastelands on a large scale. This chapter reconstructs this story by presenting the steps that led the state to become the primary agency involved in tree planting.

The reforestation as a private affair

For most of the nineteenth century, the reforestation of wastelands in Prussia was regarded primarily as a private affair. In 1766, Frederick the Great instituted a rigid supervision of forests providing punishment for clearing beyond a special budget determined by experts, but this legislation did not last long. Soon after the French revolution, unrestricted private ownership was re-established under the influence of liberal economic theories. This happened as early as 1791 in France, 1795 in Norway, 1805 in Finland, 1828 in Sweden.[9] Many among the leading Prussian Reformers attended lectures in classical economics, based on Smithian principles, by cameralists teaching in the Universities of Göttingen or Königsberg. In 1811, their reforms liberalized property law, making private forest ownership entirely free from state interference. The edict of regulation, following the abolition of serfdom in the kingdom, insured unrestricted rights to forest owners, permitting partition and conversion of forest estates. From that time, there were no obstacles to intensive timber exploitation or entire clearing and conversion into field or pasture. Furthermore, vast areas of state forests were put up for sale. This movement was also inspired by the French doctrines of Jean-Baptiste Say, whose influence was at that time very strong. Liberal thinkers regularly denounced the incapability of forestry bureaucrats and considered state monopolies over forests to be harmful. According to them, an improvement in forest management would be more readily realized if forests were privately owned. In France, about a million hectares of forest were privatized between 1807 and 1823. In Europe, only Denmark tended in the opposite direction, making state control of private forest management, established in 1681, more stringent after 1805.[10] In Prussia, Baron Karl von Stein (1757–1831), a strict follower of Smithian principles, recommended the sale of state forests most energetically. After the collapse of 1806 in the context of the Napoleonic wars, Prussia experienced a great financial distress. The need for cash demanded means to secure funds. Possible measures were widely discussed, and in 1810 to 1813, instructions for the sale of state forest property were issued. There were to be excluded from such sales only large complexes of forest, those on the seacoast, sand dunes, and river fronts, where the protection of the forest cover was needed, and those which it was desirable to maintain for industrial use.

Only the accession of Georg Ludwik Hartig (1764–1837) as chief inspector of the State forest administration in 1811 prevented further execution of this dismemberment. After 1820, however, sales continued, either for the purpose of rounding off existing state forest property or paying off use-rights lost by farmers.[11] Of 6.83 million hectares of forest in Prussia in the mid-1860s, 26.8 percent belonged to the state, 13.2 percent to the communes, 0.8 percent to diverse institutions and foundations, and 59.2 percent to private individuals.[12] Since 1820, the extent of state forests decreased from 2.41 to 2.05 million hectares. Almost 270,000 hectares had been sold only in the 1820s.[13] Since the influence of cameralists on public opinion lasted long into the century, it took a long time before the sale of state properties was stopped and finally reversed. The state foresters naturally were for retention of the existing state properties, but had little influence on public decision. As early as 1821, a Hessian forester, Georg Wilhelm von Wedekind (1796–1856), advocated the theory which later became a practice, that the state should not only retain, but also increase its present forest property by purchase of all absolute forest soil for the purpose of reforestation.[14] Among foresters, Friedrich Pfeil (1783–1859) alone was found with the cameralists on the opposite side in 1816, but, by 1834, as the director of the Forest Academy in Eberswalde, he had entirely gone over to the side of the advocates of state forest.[15] Finally, the leading political economists took a decided stand in favor of state property. Holder of the chair of political economy at Heidelberg, Karl Heinrich Rau (1792–1870) opted for opening forests to private enterprise, but with stringent regulation.[16] Robert von Mohl (1799–1875), professor of political science and political economics at Tübingen, sided against the privatization of forests, since the temptation to harvest wood prematurely contradicted the public good that would come from allowing the trees to mature. According to him, extensive regulation of wood production was essential to ensure an adequate supply at the national level.[17] For Wilhelm Roscher (1817–1894), a professor at Leipzig University, state ownership of forests had to be held since the "eternal personality of the state" was most fitting for the long growth periods of trees.[18]

In the meantime, the freedom of private forest ownership led not only to forest dismemberment but also to careless clearing. In many parts of Prussia, extensive forest devastation took place during the first decades of the nineteenth century. In order to be relieved of the privileges that the peasants had on woodlands before the abolition of serfdom, parts of these were transferred to them and almost always converted into farming lands, meadows, and pasture grounds. On poor soils, the result was sparsely vegetated land with shallow and in other cases barren land. After the fertility stored up by the forest had been exhausted by excessive removal of wood and litter, agriculture became unprofitable. Accounts from Northern Prussia describe the harsh conditions and the poverty of the poor people living on extensive sand tracts and large stretches of moor and marshland.[19] Degraded sites suffered from frequent frosts, high insolation, and strong winds and were gradually transformed into barren and almost treeless heathlands unable to furnish support to farmers. Such sterile tracts were to be found especially in West Prussia (Kashubia, Tuchel Heath), Hanover (Lüneburg Heath) and parts of Schleswig-Holstein. Where deforestation was followed by overgrazing, the soil itself could disappear. Sand drift and dust storms became common. In the 1870s, the situation prompted authorities to take active measures to restore degraded forest stands.

A partial return to state supervision of private forests

A division of opinions existed regarding the supervision of private forests by the state. Various local regulations, suited to particular districts, had, from time to time, been established, yet without satisfactory result. The political economists were inclined to reduce, the foresters to increase supervision, except Pfeil in his earlier writings. The main incentive urged by the earlier advocates of supervision was the fear of a timber shortage. This argument vanished, however, with the development of imports in the second half of the 1860s, and was then supplanted by the argument of the environmental protective functions of woodlands, especially in mountains. In the 1870s, the belief that absolute freedom of property rights in the forest was not in harmony with good political economy began to prevail.[20] In some European countries, special legislation on the management of protection forests was adopted. By passing a rigorous forest act on December 3, 1852, Austria became the leading country in this respect; France followed in 1860, Switzerland in 1876, and Italy in 1877. In Prussia, the issue was raised in 1868 and a bill submitted before the Chamber of Deputies. It resulted in the Forest Protection Law of July 6, 1875 which, for the first time, introduced some restrictions in the use of forests by private owners.

According to the law, the clearing of woodlands had to be prevented when neighboring properties could become covered by sand, exposed to the action of floods, storms, or menaced by landslides and rockfalls. In such cases, owners had to establish a "protective forest" (*Schutzwald*), where any clearing was forbidden. If their income was in any way curtailed, they were to receive indemnifications paid by those whom the devastation of forests would endanger. The latter also had to pay the costs for planting. Formally, the law introduced state intervention in forestry, but practically, most private forests remained outside of its scope. Since the law exerted restraining influence only where the evident results of clearing were seriously dangerous, a really proper basis or argument for a general protective legislation was still absent. This approach, which allowed the big landowners to exploit forests as they wanted, both satisfied the Prussian landed elite and fitted well with the liberal philosophy which continued to prevail among large circles of the Prussian administration. Karl Donner (1832–1912), who became head of the Forest Department in the Ministry of Agriculture in 1885, asserted that the duty of the state was "to subject the management of all forests to its inspection and control," but "only so far as may be necessary to obviate the dangers which an unrestrained utilization of the forest by its owners threatens to excite." For him, the rights of property were to be respected "to the utmost consistently."[21]

The law of 1875 also introduced "forest associations" (*Waldgenossenschaften*) which aimed at encouraging small adjacent woodlots, belonging to different owners, to be managed under one direction in order to take advantage of economies of scales. The institution of such an association could be demanded by the owners themselves, by the communes, the authorities, or the country police. Where property was much subdivided, and reforestation peculiarly difficult for private individuals owing to want of knowledge and means, such societies were formed, as in Westphalia, Hanover, Schleswig-Holstein. Up to 1893, twenty-six associations were constituted, covering 2,262 hectares. Half of this area was located in the district of Osnabrück.[22] In regard to reforestation, forest associations failed to be effective. The areas planted by them were disappointingly small in comparison with the vastness of the problem. At the beginning of the 1890s, Adam Schwappach (1851–1932), a senior forester working at the Academy of Forestry and the Forest Research Center in Eberswalde, affirmed that "legal compulsion" had proved to be inadequate to secure the planting of forests and "correspond[ed] no longer to modern views."[23] He therefore advocated for a higher level of state involvement, through direct financial support to private owners. This policy already existed in practice since the middle of the nineteenth century and needed to be intensified.

State support to private reforestation efforts

Reforestation required massive initial investment, generally beyond the means of the landowners, and a lifetime before any returns can be expected. In thinly populated and poor areas, no individual effort was likely to cope with such effort. State support for tree planting took diverse forms. For example, supplies of seed and plant material were granted, or part of the cost of planting was borne by the state, or loans were given. Usually, these methods were used at the same time in order to maximize the chances of success. On May 13, 1879, the Prussian Chamber of Deputies passed a law creating special credit institutions or agricultural banks in order to promote the cultivation of the soil, including the reforestation of wastelands. These banks gave loans at a moderate rate of interest (3.5 to 4.5 percent) in order to ease the burden of replanting.[24] Also, single subsidies were granted for the turning of large wastelands into forest grounds. In districts where the reforestation of private or communal land seemed especially desirable, the state or local administrative bodies paid from one-half to two-thirds of the replanting cost. It only went beyond this amount where the public interest required urgent measures, as in the case of protective forests. Other grants were paid after the planting occurred and were intended to cover the main costs of maintaining the forest in its early years. Private individuals were not only subsidized with money but were also in a position to obtain seedlings and plants on a cooperative basis, either gratis or at much reduced prices. The amount of young trees distributed by the state nurseries every year grew considerably during the period, from about 460,000 plants on average between 1870 and 1882, to 32 million in 1893.[25] If private owners were not sufficiently well off to secure experts at their own expense, trained foresters were made available for help. Their task consisted in the first place in giving advice after an inspection on the spot and watching over the correct and successful treatment of the woods.

An outlay was annually provided in the state budget for the purpose of encouraging reforestation on private lands. During the years 1881–1887, the grants distributed amounted to 426,543 marks. The largest part of this sum had been spent in the Western provinces of the Kingdom, especially in the Rhineland low mountain range and in the heathlands of Hanover and Schleswig-Holstein (see Figure 2.1).

During this period, three neighboring districts in Rhenish Prussia (Coblenz, Trier, and Aix-la-Chapelle) received more than 50 percent of the total amount granted. Almost 220,000 marks were spent in the Eifel Mountains, severely affected by deforestation due to overgrazing and massive production of charcoal for iron smelting. After 1850, the barren mountain slopes belonging to communities and private landowners were reforested with spruce at the expense of the state.[26] About 1 million marks were granted by 1892. These efforts resulted in the planting of 15,716 hectares of woodland and the restoration of forestry potential in the region.[27]

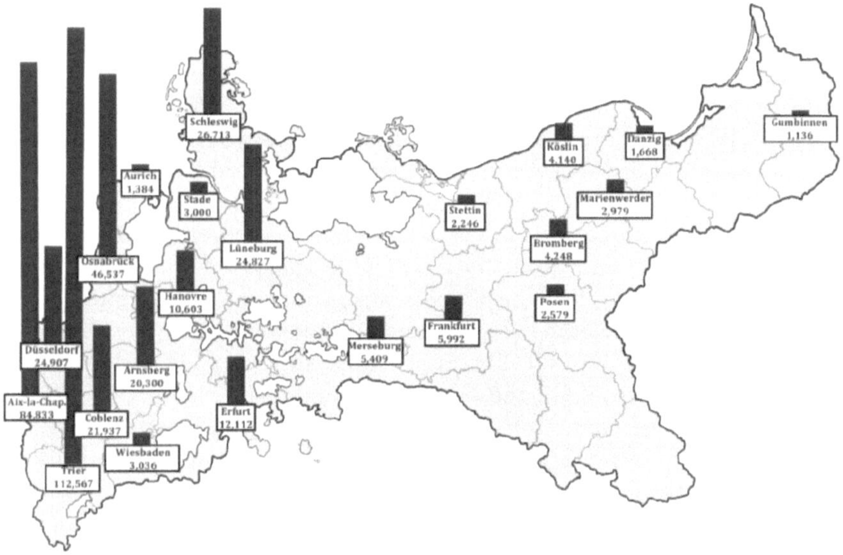

Figure 2.1 State grants allocated for reforestation on private lands (by district, in marks) (1881–1887).

Source: Ministerium für Landwirtschaft, Domänen und Forsten, *Preussens Landwirtschaftliche Verwaltung* (Berlin: Parey, 1885), 356–357 and vol. 1 (Berlin: Parey, 1888), 259–260.

State-subsidized reforestation on marginal farmland also took place in regions covered with heaths and bogs in Northwest Prussia. In Schleswig-Holstein, the State Forest administration made advances for buying plants and gave prizes for successful plantings. Within the years 1876 to 1892, private individuals afforested 3,956 hectares.[28] In the province of Hanover, a special Forest Loan-Fund was created by an Act of the Provincial Legislature on October 3, 1877. On October 16, 1880, it was enacted that associations formed for tree planting should be able to borrow funds at 2 percent interest. The amount to be borrowed was limited to the sum expended in forming the plantations. Besides, the management of the new forest was subject to the approval of the State Department of Forestry until the debt was extinguished.[29] Until 1897, the state supplied 600,000 marks.[30] Forestry officials like Georg Quaet-Faslem (1845–1919), a member of the provincial government who became later the president of the Northwest German Forest Association, actively promoted the use of the reforestation fund by local owners.[31] Large sums were loaned by means of which about 6,568 hectares of barren land were converted to forest by 1897. Approximately 75 percent of the planting operations were carried out by private individuals, the rest by communes or forest associations.[32]

Stimulation to private reforestation continued to be carried out during the following years. Since 1895, the former agricultural societies in Prussia had been organized as Chambers of Agriculture. As the official representatives of agriculture and forestry in their districts, they had means at their disposal, partly assigned by the state or provincial authorities, partly derived from contributions levied upon members. Various Chambers of Agriculture (that of Brandenburg being the first, in 1899), formed forestry sections in order to promote forest culture and reforestation of wastelands in particular. This promotional policy found expression in the institution of information bureaus, which provided technical advice and working plans. In addition, forest advisers gave lectures, instituting courses for the staffs of forest owners, and, if desired, arranging for the cooperative purchase of good seeds and plants, facilitating the sale of wood by publication of current prices. Despite some success, state support for private reforestation was not always adequate to the need. On one hand, many owners preferred to renounce the subsidy rather than subject their forest to the supervision of officials. On the other, after sums had been granted, there was no security of continuous good management from the time that the forest was formed until matured timber was placed upon the market. Around 1890, the hopes that were based on the success of individual efforts were not completely realized and although state authorities retained control of private action, this could not always be exercised, and the necessity for strengthening the state holdings became apparent.

The purchase of private land for state-led reforestation

Despite state assistance with advice, tools, seed, plants, and money, private owners had achieved little. In the 1880s, it became a common claim that only state bodies could undertake the reclamation of wasteland and the establishment of big heath plantations. The underlying argument was that forestry, because of the long growth period of trees, could not regulate itself via the market and the interests of private owners of woodlands and required intervention from the top. After 1880, such intervention was made possible by the rapid growth in government revenues (see Figure 2.2).

At the turn of the century, the evolving power of the Prussian state was better able than at any previous time to enforce appropriations of mismanaged woodlands and impoverished farms for reforestation. Political changes also accelerated the change. After the influence of liberal bureaucrats who had been prominent in the 1860s and 1870s had been reduced by Bismarck, civil servants in the central ministries sympathized with the idea that the state had to intervene to compensate for undesired social degradation.[33] After that time, the tendency was not only to maintain the state forests, but to extend their area by purchase of wastelands or by exchange for agricultural grounds. This agenda encouraged an abdication of individual involvement in connection with the bureaucracy's claim to pursue common aspirations for the good of the country. In this vision,

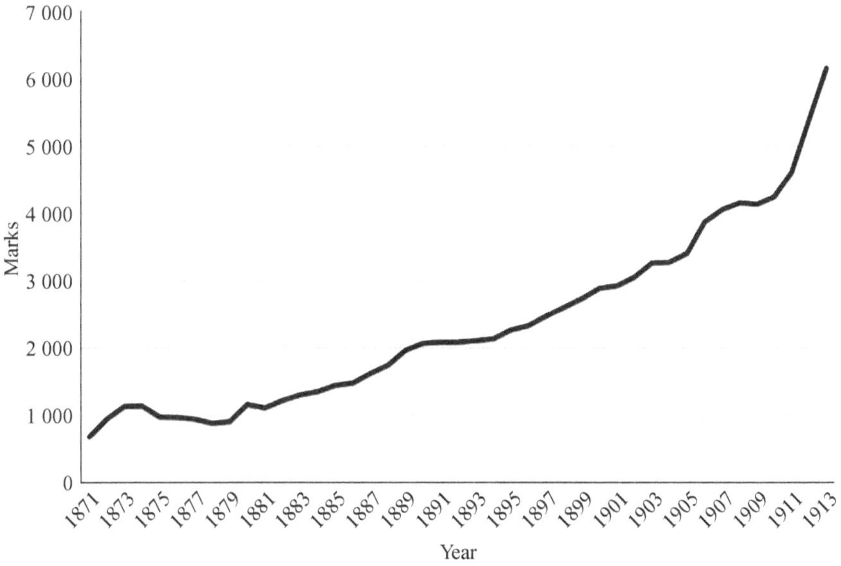

Figure 2.2 Central government revenues in Prussia in Mio. Marks (1871–1913).

Source: Peter M. Prochnow, *Staat im Wachstum. Versuch einer finanzwirtschaftlichen Analyse der preußischen Haushaltsrechnungen 1871–1913.* (PhD Thesis: Westfälische Wilhelms – Universität zu Münster, 1977), Tabellenband, p. 6.

the state was conceived as a householder, supplied with special knowledge and able to reach a balance between arable and woodlands across the Kingdom. In the long term, tree planting by the state was expected to generate a surplus, prevent environmental degradation, and alleviate poverty.

In some places, this policy had already been implemented since the middle of the nineteenth century, but on a very small scale. Private tracts were bought or exchanged to round off the state holdings but also for planting trees. After 1846, some funding was available for this purpose. The sums actually allocated, however, were derisory: up to 1871, only 1 million thalers was used for land purchase, far less than the 6.9 million spent for dissolving use-rights on state lands through cash payments to peasants.[34] This financial involvement nevertheless allowed some state foresters to make progress. During the 1870s to 1880s, several local campaigns using reforestation to improve local agricultural conditions began to gain attention. In the Lüneburg Heath, the director of the Hanoverian State forest department, Heinrich Christian Burckhardt (1811–1879), led the purchase of extensive stretches of soil unfit for cultivation. In 1872, he presented the results achieved and claimed the necessity to make further acquisitions.[35] In subsequent years, many farmers sold their properties, which were then planted with pine. As a result, the area of heath was drastically reduced.[36] State holdings also extended in West Prussia. In the late 1880s, Reinhold Schütte (1830–1897), a chief forester in the Tuchel Heath, produced a short book documenting the success of reforestation efforts since the mid-nineteenth century.[37] According to Schütte, the introduction of tree planting had transformed the region and was beginning to create a productive landscape. At the beginning of the 1890s, his proposal to increase purchases received approval from senior forestry officials in Berlin.[38]

Beginning in 1871, the Prussian Chamber of Deputies provided an ordinary budget allotment for the purchase of wastelands in order to make the reforestation effort more systematic.[39] As means became more plentiful, the budgets were increased, not merely for the purchase of lands but for planting operations as well. In 1872, 375,000 marks were allocated, in 1873, 525,000 marks, and in 1874 more than 1 million.[40] In 1877, the deputies decided to proceed more quickly. The parliamentarian and director of the Hanoverian Academy of Forestry, August Bernhardt (1831–1879), demanded an acceleration of the purchase tempo and the conversion of some public estates unfit for cultivation into woodlands.[41] In subsequent years, the purchase of degraded forests, waste and heath lands was steadily carried out. In 1882, the budget for annual appropriations was increased to 2 million marks. Planting enthusiasts increasingly used numbers to encourage massive purchase of wastelands by the state, making deforestation a matter of grave national concern and presenting tree planting as a common good providing various collateral benefits for society. Through their writings, they both created a sense of urgency and demonstrated success, as indicated by some interim or final tally of trees planted.[42] Up to 1892, almost 135,000 hectares were taken in hand at a cost of about 22.4 million marks (on average 167 marks per hectare, including timber).

At this time, state foresters aimed at joining those other great corps of engineers—civil, hydraulic, and mining—that were so influential in developing the country. Concerned about their public image, they promoted themselves as responsible stewards of the environment. In forestry journals, state foresters employed a crisis narrative to legitimize state intervention in the form of reforestation. The alleged crisis involved timber shortage, erosion, flooding, and various other resource and environmental problems. At the turn of the 1880s, such propaganda work was echoed by an influential group of agronomists and political economists whose many writings on agriculture and land reclamation opened a bridgehead to the reforestation movement and helped to create a broad consensus around this need. A growing number of academics like the agronomist Max Sering (1857–1939) or the businessman and politician Alfred Hugenberg (1865–1951) demanded a large-scale, centrally organized and state-controlled policy of internal colonization in Germany.[43] In their view, the state had to assist the creation of a rural middle class who would act as a bulwark against the appeal of the socialists. Tree planting had a crucial role to play as a means of harnessing nature's productive potential and maximizing the economic productivity of marginal lands where colonists had to be settled. Agronomists and rural reformers repeatedly stressed the importance of forests for the peasant economy, since farmers depended on woodlands for fuel for building materials and supplements for their diets.

One of the underlying influences on state involvement was a desire to make productive use of otherwise unproductive land. As the population grew and the use of wood increased, greater attention was paid to the formation of new forests in order to make Germany self-sufficient in timber. In 1892, August Pflug published interesting data about the topic in a book documenting the reforestation efforts already made and the remaining possibilities for expending the forest area.[44] One year later, the State Agricultural Commission (*Landes-Oekonomie-Kollegium*) asked for a census to be made of poor meadows, rough pastures, waste and sterile land physically suitable for growing trees across the kingdom. As a result, it was established that Prussia owned 3.2 million hectares of such grounds, of which 587,000 could be reforested. Most of them were located in Hanover, Schleswig-Holstein and West Prussia (see Figure 2.3).

At the end of the nineteenth century, Germany was rapidly approaching a shortage, if not actual dearth, in its supply of coniferous timber. In spite of the fact that the German Empire succeeded in producing the largest amount of wood per hectare in Europe (about 3.53 cubic meters in 1900), it was unable to supply its own needs for timber.[45] After 1863 the imports in

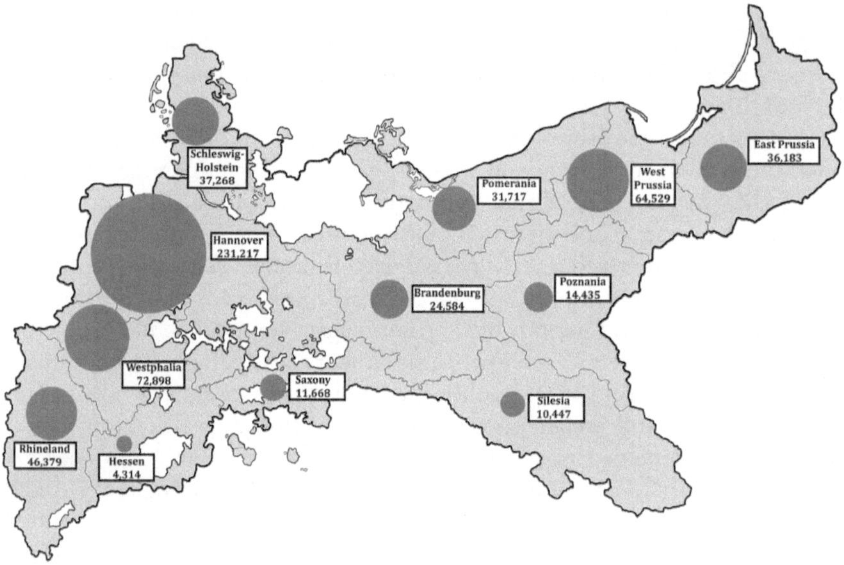

Figure 2.3 Extent of wastelands suitable for reforestation (by province, in hectares) (1893).

Source: Bernhard Danckelmann, "Wirtschaftliche und wirtschaftspolitische Rückblicke aus landwirtschaftlichen, forstlichen und gewerblichen Kreisen auf Forstwesen und Jagd des Jahres 1893 in Preußen," *Zeitschrift für Forst- und Jagdwesen* 27, no. 5 (1895): 249–281, 253.

logs, sawed and hewn timber, mostly from Russia and Austria–Hungary, have exceeded the exports. The difference between them more than doubled at the turn of the century, rising from 5.5 million cubic meters in 1890 to 11.6 million cubic meters in 1910.[46]

Although Prussia held 59.1 percent of the forest area of the German Empire, it furnished only 47.6 percent of the total production of timber over 2.75 inches (*Derbholz*), and only 48.3 percent of timber production; in each case it was less than half.[47] For Ketzler, reforestation had become "the fundamental question of the Prussian forest policy for the present and the foreseeable future."[48] The most optimistic experts claimed that reforestation could remedy this situation, at least partly. According to Richard Zimmermann, the extent of the state forest (about 2.65 million hectares in 1893) could be doubled through a large program of reforestation; the state would then be in possession of half of the Prussian woodland area.[49] All these plans seemed to capture the imagination of state officials in Berlin at the beginning of the 1890s. The reforestation message was, for instance, enthusiastically embraced by Gustav von dem Borne (1832–1916), a senior official who, after an early career in Hanover, became chief of the forest bureau in the Ministry of Agriculture in 1891. Borne sought to intensify land acquisition in West Prussia and, in 1892, he published an article to this effect in a prominent forestry journal.[50] After 1893, all the receipts from sales of public domains were set aside to secure wasteland and forest properties. Up to 1910, the state forest administration acquired 334,872 hectares of land. The area annually taken in hand varied from 3,777 hectares (1870–1890) to 12,967 hectares (1890–1910) on average. After 1890, efforts clearly focused on the eastern provinces of the Kingdom (see Figures 2.4a and 2.4b).

Larger and smaller tracts were bought and after the sale or exchange of parts suitable for agriculture or pasture, the soils fit for forming compact woodlands were planted by the administration. State plantations consisted mainly of Scots pine or spruce, according to the soil. Most of the state-owned forests were regular timber forest where the final crop was intended to furnish saw timber, and every particular parcel was supposed to be stocked with trees (mostly pine and spruce) of nearly the same age.[51] Such plantation forestry was based on a productivist model aimed at growing a particular type of forest crop for the nation's needs. Since this program assumed a permanent state bureaucracy to manage forests indefinitely, it strengthened the social role of state foresters.

Special attention was given to reforestation in the eastern parts of the Kingdom, especially in the province of West Prussia, where the need was the greatest. In the sandy plains, the rural population consisted mostly of impoverished Kashubians and Polish farmers having no means for planting

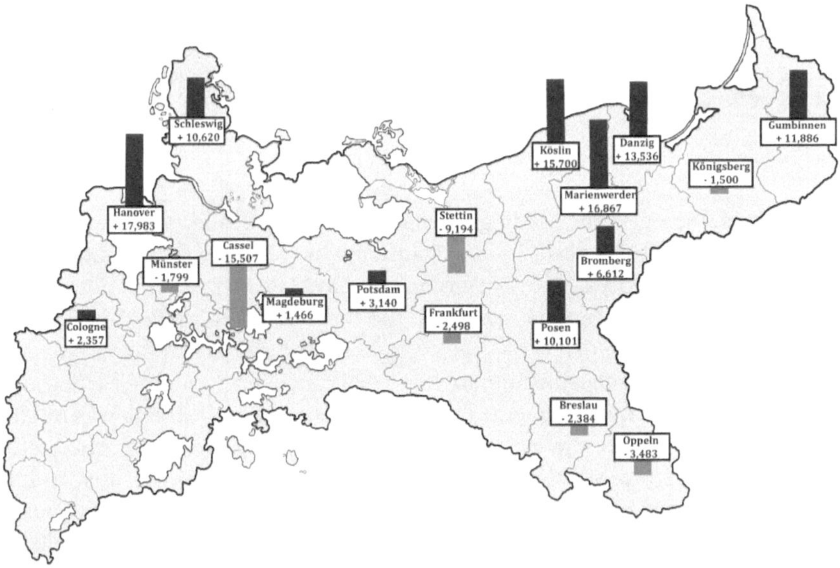

Figure 2.4a Gains and losses in the extent of the State Forests (by district, in hectares) (1870–1890).

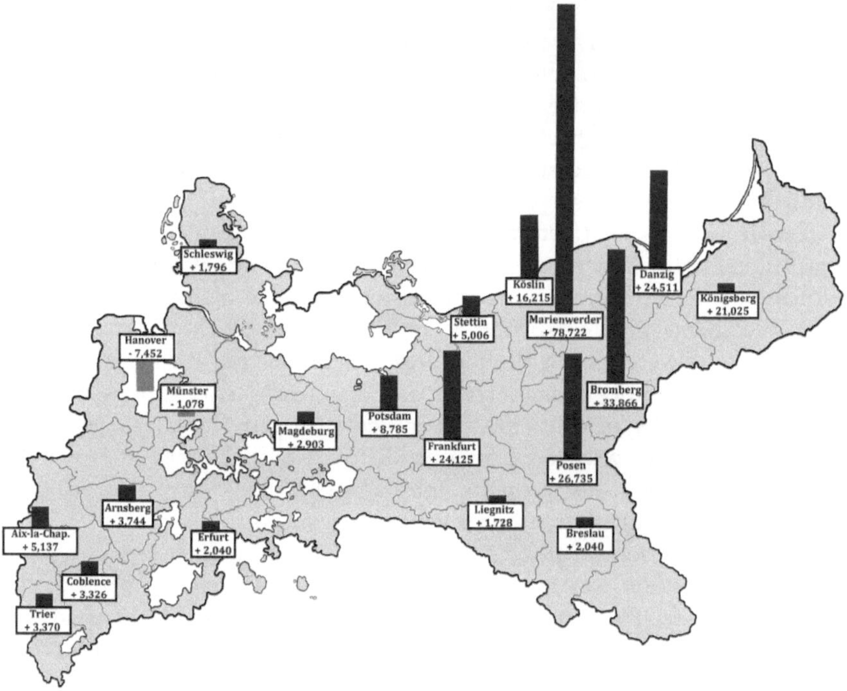

Figure 2.4b Gains and losses in the extent of the State Forests (by district, in hectares) (1890–1910 (to 1905 for Königsberg)).

Sources: Otto von Hagen, *Die forstlichen Verhältnisse Preussens*, vol. 2 (Berlin: Springer, 1884), 41–44; *Amtliche Mitteilungen aus der Abteilung für Forsten des Königlich Preußischen Ministeriums für Landwirtschaft, Domänen und Forsten. 1910* (Berlin: Springer, 1912), 38.

Note
Only significant changes (greater than 1000 ha) are represented.

operations. Since their holdings were mostly small and scattered, much difficulty was experienced in amalgamating large enough properties for reforestation purposes.[52] Forestry officials also doubted the effectiveness of state-supported planting operations in the local cultural context. In 1889, a report from the Ministry of Agriculture about the sandy areas around Karthaus presented the Kashubian "national character" as a major obstacle to local economic development. Located in Pomerelia, Kashubia was inhabited by a Slavonic population, most closely related to the Poles, who traditionally lived from subsistence farming, fishing, and animal husbandry. According to the report's author, these poor people were lacking "energy and intelligence" and their "indolence" would make them incapable of using rationally any financial help for planting.[53] More generally, the state forest administration argued that reforestation had to be managed differently in the East: here it needed to be state-led rather than privately funded as in the heaths of Hanover or the hills of Rhenish Prussia. With only 12,818 marks allocated between 1881 and 1887, the provinces of Poznania, West Prussia, and East Prussia obtained less than 3 percent of the total amount granted throughout the Kingdom (see Figure 2.1).

After the mid-1880s, West Prussia—as well as Masuria in East Prussia and the northern part of the neighboring province of Poznania—became laboratories of large-scale state-led reforestation in Prussia. From 1887 onward, the Royal General Commission sitting in Bromberg began to make larger investments in the reforestation of wasteland in coordination with the Ministry of Agriculture in Berlin.[54] On January 28, 1892, the Minister issued a directive setting the way to proceed with tree planting on wastes.[55] Efforts were made to reacquire barren lands in Kashubia. Up to 1892, an additional 16,200 hectares of degraded forests and wastes were bought by the State forest administration in the circles of Schlochau, Konitz, and Berent.[56] A comparison of two forest maps drawn by the administration at different moments gives an indication as to the rapid progress of state holdings in this area (see Figure 2.5).

In most cases, the purchased lands were former property of indebted farmers. Land consolidation proceedings (*Zusammenlegungsverfahren*) were used by a special commission in Konitz in order to speed up the transfer of ownership.[57] While private owners have continued to decrease their forest

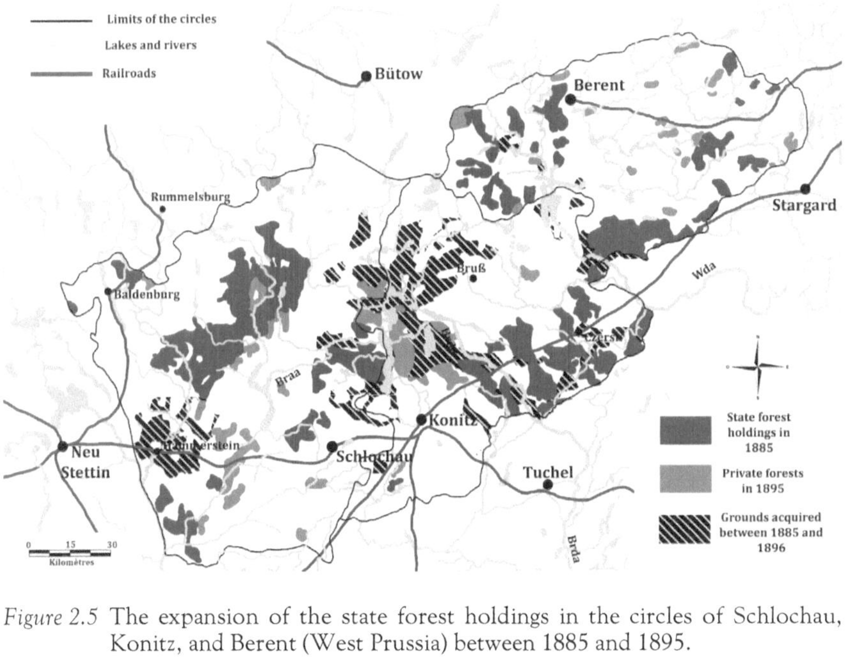

Figure 2.5 The expansion of the state forest holdings in the circles of Schlochau, Konitz, and Berent (West Prussia) between 1885 and 1895.

Sources: Ministerium der Öffentlichen Arbeiten, *Uebersichts-Karte von den Waldungen Preussens* (Berlin: Julius Springer, 1887, 1897).

area, the increases in State forest were largely secured by planting up wastelands. Up to 1906, nearly 45,000 hectares had been reforested in West Prussia alone, partly by sowing, partly by planting; hundreds of mobile sand dunes were stabilized.[58]

Reforestation and the authoritarian practices of the Prussian administration

In most cases, state-led reforestation went hand in hand with authoritarian methods. The belief in the state as planter rested on new scientific knowledge about environmental processes, which could be seen and regulated only "from the top." In the eyes of state foresters, heath people had proved incapable of managing the forests, preserving its soil, or conserving its vegetative cover. By acquiring land for reforestation, the state affirmed itself as the new gardener of the forest. Such conception of intervention in the social realm was not specific to foresters. Having investigated the ideas and policies of Prussian officials involved in the construction of the Welfare State, Hermann Beck identifies a "mentality of benevolence and patriarchal concern, pervaded by authoritarian streaks, that was unique in nineteenth-century Europe."[59] According to Beck, Prussian interventionism derived its legitimation from the bureaucracy's claim to work for the good of the population. But regarding forests, such a mission was often misunderstood and gave rise to many conflicts between foresters and forest users. As pointed out by James C. Scott in his book devoted to "high modernist schemes," the invention of scientific forestry in late eighteenth-century Prussia and Saxony was based on a reduction of the real forest complexities to a single number, the revenue yield. From that time, foresters developed a tendency to control and simplify social and ecological contexts, in order to increase output efficiency in the exploitation of forests resources.[60] As large areas of land came under the jurisdiction of the Prussian state forest administration in order to be reforested, earlier perceptions of the forest as the source of a multiplicity of goods and services were over.

In Rhineland—a territory which had passed under the domination of Prussia in 1815—local farmers resented the conversion of their traditional beech and oak coppicewoods (which provided fuel wood, stall litter, and pig feed) into high coniferous forests (which yielded only industrial timber) as the imposition of a "foreign" forest culture. The tracts were replanted mostly with Norway spruce (*Picea abies*), not with the "naturally" occurring deciduous tree species of the Eifel and Hunsrück. Villagers resisted state foresters' policies, imposed from above, to keep their heath and underbrush from reforestation with coniferous trees.[61] They refused to plant what they called a "Prussian tree" (*Prussenbaum* or *Prüsseboom* in the local dialect) and protested against reforestation. In the eyes of state foresters, such resistance proved that the "backward" local farmers were not conscious of their own interests. At first, they tried to "educate" the peasants, but quickly turned to commands and threats, when they proved unwilling to learn.[62] In some cases, the newly planted trees were uprooted by hostile farmers, and soldiers were required to protect plantations.[63]

In West Prussia, the transformation of "unproductive" landscapes into plantations of fast-growing softwood trees also led to numerous conflicts with the population. The reforestation with conifers was the best way of making it impossible for farmers to use the forest as pasture. In order to preserve plants and young trees on newly reforested land, Prussian officials enforced regulations excluding grazing livestock from the forest. Local residents tended to ignore such appeals and continued to turn their cattle onto state forests grounds, regardless of instructions.

With the progress of reforestation, smallholding agricultural and pastoral producers increasingly lost control over how the forests were managed. In Kashubia, state-led reforestation became a symbol of land dispossession, as was earlier the case in Rhineland. Such feeling found expression in the publications of the Young Kashubians, a regionalist organization which aimed at awakening ethnic self-awareness and Polish patriotism among the Kashubian people. In 1910, the poet and journalist Aleksander Majkowski (1876–1938), the most important figure in the movement, staged his own discursive struggle against landscape and demographic changes caused by reforestation. Deploring the agricultural decline and the ruin of Kashubian villages in the reforested areas, he spoke of the Prussian State forest as a "green expropriator" driving the Kashubian out of their land.[64]

As a matter of fact, Prussian officials did not want to depopulate the region, on the contrary.[65] Reforestation efforts required not only a bureaucracy of foresters but also thousands of workers to plant, harvest, and protect forests. Usually, the state forest administration employed local people on a contract basis or seasonal workers on a day-labor basis. But when labor was scarce, which was very common in these sparsely populated areas, forestry officials often made use of authoritarian methods in order to implement reforestation projects. That was particularly the case during the spring, when planting operations demanded a lot of workers. In the province of Hanover, regional authorities resorted to a new and original method, by using vagabonds, tramps, and prisoners for forest culture. In this manner about 3,600 hectares were planted in the years from 1876 to 1888.[66] In Oerrel, near Lintzel, the plan of reforestation was started in 1876 with the purchase of two estates unsuitable for agriculture. Since the reforestation areas were located in a thinly populated district, it became evident that the local supply of labor would be insufficient. This difficulty was partly overcome by the importation of colonists, but the chief labor supply was derivable from a prison erected at Oerrel, and which was capable of containing 150 inmates.[67] Such practices were also prevalent in West Prussia. In the Hel Peninsula, a long sandbar separating the Bay of Puck from the open Baltic Sea, local fishermen did not intend to work in the forests. Facing a serious lack of labor force during the 1890s, forest inspectorates brought

Kashubian female workers from the Karthaus area, but this operation was expensive and did not bring the expected success. From 1898 onwards, prisoners from the penitentiary of Danzig were forced to work on the plantations.[68] In the Konitz region, the State forest administration occasionally resorted to children who were released from their classes for a couple of weeks in order to help with planting trees.[69]

Conclusion

On the eve of the First World War, the state was undoubtedly the primary agency involved in tree planting for both timber and non-commercial purposes in Prussia. During the previous three decades, efforts to replenish the forest resource occurred largely through reforestation by the state. In 1910, the state forest administration employed a total of 5,030 officials directly managing 3.0 million hectares of land and supervising large areas of private grounds being reforested with the help of state grants.[70] Standing at the head of this powerful administration, the Prussian state proudly promoted itself as a tree planter and as partner to other planting groups. Such an image was anchored in the reality of a rapidly changing landscape. As a result of various financial and logistical efforts from the state, Prussia experienced first a stabilization, then an increase in forest cover. Prussian involvement in reforestation exerted an empire-wide impact. In 1913, the woodlands of the German Empire covered 14.2 million hectares, 300,000 hectares more than in the early 1870s.[71] The reversal of the long-established trends of deforestation reflected the triumph of the Prussian state power and can be interpreted as evidence for an "environmental statehood" in the sense defined by Ioris. By intervening directly in the reforestation movement, either by purchasing and reforesting wastelands or by sponsoring tree planting on private holdings, the Prussian state acted as a Hobbesian Leviathan, saving the environment from society overexploiting forests. At the same time, the methods used by state officials in the reforestation areas manifested Prussian typical conceptions of authoritarian rule. In order to preserve the nation's property, tree plantations were placed under rigid regulations regarding forest use and the organization of labor. But on the other hand, the liberal legislation allowing owners to use their forest as they wanted was never repealed. In this respect, the Prussian way to deal with deforestation and to reforest wastelands was a strange combination of liberalism and interventionism.

Notes

1 Hermann Beck, *The Origins of the Authoritarian Welfare State in Prussia: Conservatives, Bureaucracy, and the Social Question; 1815–70* (Ann Arbor, MI: University of Michigan Press, 2000).
2 Bernhard E. Fernow, *A Brief History of Forestry in Europe, the United States and Other Countries* (Toronto: Toronto University Press, 1911), 143–144.

3 August Meitzen, *Der Boden und die landwirthschaftlichen Verhältnisse des Preussischen Staates nach dem Gebietsumfange vor 1866* (Berlin: Wiegandt & Hempel, 1868).
4 Joachim Radkau, "Wood and Forestry in German History: In Quest of an Environmental Approach," *Environment and History* 2, no. 1 (1996): 63–76, 72.
5 Henry E. Lowood, "The Calculating Forester: Quantification, Cameral Science, and the Emergence of Scientific Forestry Management in Germany," in T. Frängsmyr, J. L. Heilbron, and R. E. Rider (eds.), *The Quantifying Spirit in the 18th Century* (Berkeley, CA: University of California Press, 1990), 315–342.
6 A. Howard Gron, "The Economic Foundations of Forest Politics," *Unasylva* 1, no. 3 (1947).
7 Antonio Ioris, *The Political Ecology of the State: The Basis and the Evolution of Environmental Statehood* (London: Routledge, 2014), VII.
8 Fernow, *A Brief History*, 121.
9 Gron, "The Economic Foundations."
10 Ibid.
11 Fernow, *A Brief History*, 93–94.
12 Otto von Hagen, *Die forstlichen Verhältnisse Preussens* (Berlin: Springer, vol. 2, 1884), 6.
13 "Die Waldflächen-Verhältnisse und die Aufforstung von Oedländereien in der Preußischen Monarchie," in *Sammlung sämmtlicher Drucksachen des Hause der Abgeordneten aus der XI. Legislatur-Periode. III Session* (Berlin: W. Moeser, 1873), 5.
14 Wilhelm von Wedekind, *Versuch einer Forstverfassung im Geiste der Zeit* (Leipzig, 1821).
15 Wilhelm Pfeil, *Über die Ursachen des schlechten Zustandes der Forsten* (Züllichau: Dammann, 1816), 79; Wilhelm Pfeil, *Die Forstpolizeigesetze Deutschlands und Frankreichs* (Berlin: Veit, 1834), 173.
16 Karl Heinrich Rau, *Lehrbuch der politischen Oekonomie* (Heidelberg: Carl Winter, 2nd ed., vol. 2, 1839), 232–233.
17 Robert von Mohl, *Polizeiwissenschaft nach den Grundsätzen des Rechtsstaates* (Tübingen: Laupp, 2nd ed., vol. 2, 1844), 232–237, 247–248.
18 Wilhelm Roscher, *Ein nationalökonomisches Hauptprinzip der Forstwissenschaft* (Leipzig: Hirzel, 1854), 22.
19 Ministerium für Landwirtschaft, Domänen und Forsten, "Aufnahme über die allgemeine Lage der Landwirthschaft in einem Gemeindebezirk wohl des ärmsten Kreises von Westpreussen mit leichtem Sandboden. Herbst 1889," in *Ermittelungen über die allgemeine Lage der Landwirthschaft in Preussen. Aufgenommen im Jahre 1888/89* (Berlin: Parey, vol. 2, 1891), 21–38.
20 Fernow, *A Brief History*, 143–144.
21 Quoted after: Gifford Pinchot, "Government Forestry Abroad," *Publications of the American Economic Association* 6, no. 3 (1891): 7–54, 21.
22 Hagen, *Die forstlichen Verhältnisse*, 39.
23 Adam Schwappach, "Information as to the Assistance, by Means of Loans, Inspection, Remission of Taxation, etc., that Foreign States give to Planters, and the Encouragement They Offer to the Reforestation of Waste Lands," *Minutes of Evidence, Committee on British Forestry* (London: Wyman and Sons, 1903), 210–211 (appendix XXV).
24 Ibid.
25 Ibid. Ernst Arndt, *Die Privatforstwirthschaft in Preussen* (Berlin: Springer, 1889), 156.
26 Irmund Wenzel, *Ödlandentstehung und Wiederaufforstung in der Zentraleifel* (Bonn: Dümmlers, 1962).
27 Max Endres, *Handbuch der Forstpolitik, mit besonderer Berücksichtigung der Gesetzgebung und Statistik* (Berlin: Springer, 2nd ed., 1905), 387.

28 Hahn, "Die Aufforstungen in Schleswig-Holstein," *Zeitschrift fur Forst- und Jagdwesen* 25, no. 5 (1893): 249–283, 254.
29 "Forest Culture in Hanover," *Science* 16, no. 392 (1890): 78.
30 Quaet-Faslem, "Die Aufforstungsbestrebungen der Hannoverschen Provinzialverwaltung," *Zeitschrift für Forst- und Jagdwesen* 28, no. 1 (1896): 32–46, 33.
31 Quaet-Faslem, "Zur Provinzial-Aufforstung," *Land- und fortwirthschaftliches Vereinsblatt für das Fürstenthum Lüneburg* 4 (1881), 25–26; Quaet-Faslem, "Über die Aufforstungs-Bestrebungen der Hannoverschen Provinzial-Verwaltung," *Forstliche Blätter* 19, no. 2 (1882): 41–47.
32 Quaet-Faslem, "Die Aufforstungs-Bestrebungen," 45.
33 Hans-Ulrich Wehler, "Der Aufstieg des organisierten Kapitalismus und Interventionsstaates in Deutschland," in H. A. Winkler (ed.), *Organisierter Kapitalismus. Voraussetzungen und Anfänge* (Göttingen: Vandenhoeck & Ruprecht, 1974), 36–57.
34 "Die Waldflächen-Verhältnisse."
35 Heinrich Christian Burckhardt, "Die Aufforstung der Haiden. Eine Studie aus Hannovers Heiden," *Aus dem Walde*, 3 (1872): 41–85.
36 "Forest Culture," 78.
37 Reinhold Schütte, *Die Tucheler Haide vornehmlich in forstlicher Beziehung* (Danzig: Bertling, 1893).
38 Jeffrey K. Wilson, "Environmental Chauvinism in the Prussian East: Forestry as a Civilizing Mission on the Ethnic Frontier, 1871–1914," *Central European History* 41, no. 1 (2008): 27–70.
39 "Die Waldflächen-Verhältnisse."
40 H. A. Reinick, *Beiträge zur Waldschutz- und Aufforstungs-Frage mit besonderer Beziehung auf die Provinz Hannover* (Hildesheim: August Lax, 1881), 16; König, "Die Entwicklung der staatlichen Forstwirtschaft in Westpreußen und ihre Beziehungen zur Landeskultur," in *Bericht über die VII. Hauptversammlung des Deutschen Forstvereins* (Berlin: Springer, 1907), 145–176, 167–168.
41 August Bernhardt, "Die Umwandlung der preussischen Staats-Domänen in Staats-Forsten," *Landwirtschaftliche Jahrbücher* 7 (1878): 1–18, 3, 8.
42 Gustav von dem Borne, "Ueber den Fortgang der Oedlandserwerbungen und Aufforstungen in den Regierungsbezirken Königsberg, Danzig und Marienwerder und über den Erfolg dieses Vorgehens der Staatsforstvewaltung," *Zeitschrift für Forst- und Jagdwesen* 28, no. 12 (1896), 699–714.
43 Alfred Hugenberg, *Innere Colonisation im Nordwesten Deutschlands* (Strassburg: Trübner, 1891); Max Sering, *Die innere Kolonisation im östlichen Deutschland* (Leipzig: Duncker & Humblot, 1893).
44 August Pflug, *Die wirtschaftliche Erschliessung öder und geringwertige Liegenschaft durch künstliche Aufforstungen. Sonderabdruck aus der Zeitschrift für die gesamte Staatswimenchaft* (Tübingen: Laupp, 1892).
45 Jawad Daheur, "La sylviculture allemande et ses 'hectares fantômes' au tournant des XIXe et XXe siècles," *Revue forestière française* 3 (2017): 227–239.
46 Endres, *Handbuch der Forstpolitik*, 586–587.
47 Raphael Zon, *The Forest Resources of the World* (Washington, DC: Government Printing Office, 1910), 42.
48 W. Ketzler, "Ueber die Aufforstung von Oedländereien," *Zeitschrift für Forst-und Jagdwesen* 15, no. 8 (1883): 426–443, 430.
49 Richard Zimmermann, "Deutschlands Holzbedarf," *Zeitschrift für die gesamte Staatswissenschaft* 50, no. 4 (1894): 573–582, 577.
50 Gustav von dem Borne, "Die Oedlands-Ankäufe und Aufforstungen der Preußischen Staatsforst-Verwaltung mit besonderer Berücksichtigung der westpreußischen Kassubei," *Zeitschrift für Forst-und Jagdwesen* 24 (1892): 393–406.

51 Bernhard E. Fernow, *Forest Policies and Forest Management in Germany and British India* (Washington, DC: Government Printing Office, 1899), 226.
52 Carl Gerlich, *Statistische Beschreibung des Schwetzer Kreises* (Schwetz: J. Bauffe, 1878), 53.
53 Ministerium für Landwirtschaft, Domänen und Forsten, "Aufnahme über die allgemeine Lage," 29.
54 König, "Die Entwicklung der staatlichen Forstwirtschaft in Westpreußen und ihre Beziehungen zur Landeskultur," in *Bericht über die VII. Hauptversammlung des Deutschen Forstvereins* (Berlin: Springer, 1907) 145–176: 167–168; Borne, "Die Oedlands-Ankäufe," 396.
55 Bernhard Danckelmann, "Aufforstung von Oedländereien," in *Jahrbuch der Preußischen Forst- und Jagdgesetzgebung und Verwaltung* (Berlin: Springer, 1892), 151–152.
56 Borne, "Die Oedlands-Ankäufe," 401.
57 Semper, "Die Ankaufspolitik der preussischen Staatsforstverwaltung in Westpreusen und Posen," *Zeitschrift für Forst- und Jagdwesen* 43, no. 2 (1911): 65–92, 85–86.
58 Tuisko Lorey and Christof Wagner, *Handbuch der Forstwissenschaft* (Tübingen, Laupp, 3rd ed., vol. 1, 1913,), 132.
59 Beck, *The Origins*, X (Preface).
60 James C. Scott, *Seeing like a State. How Certain Schemes to Improve the Human Condition Have Failed* (New Haven, CT: Yale University Press, 1998).
61 Reinhold Wacker, *Das Land an Mosel und Saar mit Eifel und Hunsrück: Strukturen und Entwicklungen 1815–1990* (Trier: Spee, 1991), 164.
62 Jonathan Sperber, *Rhineland Radicals: the Democratic Movement and the Revolution of 1848–1849* (Princeton, NJ: Princeton University Press, 1993), 75.
63 Karl Hasel, *Forstgeschichte* (Hamburg: Paul Parey, 1985), 209.
64 Aleksander Majkowski, "Zielony wywłaszczyciel [The green expropriator]," *Gryf : pismo dla spraw kaszubskich* 2, no. 6 (1910): 161–168.
65 Georg Stockfleth, "Zur Seßhaftmachung der Waldarbeiter," *Zeitschrift für Forst- und Jagdwesen* 10, no. 14 (1895): 205–208.
66 "Forest Culture," 78.
67 Somerville, "A Short Account of the State Forests of Prussia (including Hanover)," *Transactions of the Royal Scottish Arboricultural Society* 14 (1895): 140–162, 141–142.
68 Spiegel, "Hela," *Bericht des Westpreussischen Botanisch-Zoologischen Vereins* 26/27 (1905): 126–142.
69 Konrad Agahd, *Kinderarbeit und Gesetz gegen die Ausnutzung kindlicher Arbeitskraft in Deutschland: unter Berücksichtigung der Gesetzgebung des Auslandes und der Beschäftigung der Kinder in der Landwirtschaft* (Iena: Fischer, 1902), 160.
70 Hagen, *Die forstlichen Verhältnisse*, 43; *Amtliche* Mitteilungen: 29.
71 "Der Deutsche Wald," *Deutscher Reichs-Anzeiger und Königlich Preußischer Staats-Anzeiger* 35 (1872): 3–4; Lorey and Wagner, *Handbuch der Forstwissenschaft*, 52.

3 Red and green all over

Counterinsurgency and conservation in the jungles of Cold War Guatemala

Tony Andersson

Introduction

Forest conservation and anti-communist counterinsurgency share a long history in northern Guatemala's tropical lowlands, a historically isolated frontier region often compared to a Central American Amazon. There, in the province called the Petén, the Guatemalan army built a parallel state that blended scientific forest management with counterinsurgency during a long and bloody civil war (1960–1996) that pitted the army against leftist guerrillas and civilian peasants. The war reached a sanguinary peak with the army's genocide of indigenous Maya villagers in the early 1980s. It has been the subject of considerable scholarly attention, popular literature, activism by Nobel laureates, and war crimes trials that are ongoing today. Yet few accounts of that conflict have appreciated the role of woodlands in what was ostensibly a battle over farmland. From the beginning, the Guatemalan army centered the lowland forests within its national security doctrine, fusing the institutions of forest management and the violent repression of political dissent. This administrative approach to people and forests, which I call counterinsurgent environmentalism, endured even as the goals and methods of both evolved over time, and has continued to inform conservationist practice under the nominally democratic governments that followed the end of the war.

The experience of the Petén raises several important qualifications to the arguments of this volume. First, Guatemala's military regimes never saw conservation independently of their preoccupation with rural social control. For the army officers charged with eliminating an elusive and often spectral internal enemy, control over the landscape and the people on it were one and the same thing. Although indivisible, social control and conservation were not equal: the former always got priority, guiding the form and function of the latter. Second, conservation in Guatemala did not involve a monolithic state acting in the interest of (or against) the singular value of "environmental protection." Rather, it involved a struggle among rival social groups within an authoritarian regime to define what was meant by environmental degradation, who had caused it, who had the authority to design and implement remedies, and who would bear its burdens. Counter-insurgent environmentalism was just one of many competing environmental ethics in Guatemala, and examining its path to predominance helps illuminate the dynamic local, national, and international power structures in play in one of the Cold War's bloodiest "brushfire conflicts." Finally, Guatemala's experience belies the implied distinction between "Western democracies" and "authoritarian regimes" in the assertion that totalitarian states have a competitive advantage over liberal democracies when it comes to meeting the challenges of pollution, resource preservation, or climate change. The conservation institutions that the Guatemalan army built relied on substantial financial and technical support from the United States, West Germany, multilateral development agencies like the World Bank, and private organizations like the Rotary Club. They were also explicitly modeled on examples from democratic countries—above all, the Tennessee Valley Authority (TVA), an immense technocratic bureaucracy intended to sidestep the challenges of democratic decision-making.

This chapter elaborates these arguments using two distinct moments from Guatemala's civil war, one near its beginning and the other near its end. The first moment corresponds to the rise of a true military state in the 1960s, which made the vast woodlands of the Petén the centerpiece of its counterinsurgent strategy. The second moment came in the 1980s, during and after the army unleashed a genocidal scorched earth campaign against the civilian population living near areas of guerrilla activity. In both moments under discussion here, the army wielded forest conservation as a weapon against politicized peasants. What that meant for the forests and the people living in them, however, was radically different in each case. The high command's view of the forests changed in response to political rivalries within the right-wing coalition that supported it, the growth of insurgents' military and political strength, as well as scientific studies and the rise of a global environmentalist movement. The fate of Guatemala's rainforest landscapes cannot be understood through a binary framework of democracy or despotism, but only through the dynamic intersections of scientific knowledge, geopolitics, and internal power relations working at multiple scales within the authoritarian system.

The Petén, a "quintessential frontier"

In the Guatemalan popular imagination, the Petén is roughly equivalent to the Brazilian Amazon: a geographically distinct region of lowland tropical forests. It is hot, densely forested, contains some of the richest stands of mature mahogany trees in the world, and is drenched in rain for half of the year. Bounded on the north and west by Mexico, and by Belize to the east, the province encompasses a third of the national territory (see Figure 3.1). Its geography made it easily accessible by river or mule trail from Mexico and Belize, but remote from the Guatemalan highlands, where most

Figure 3.1 Map of Petén in regional context.

Guatemalans lived. Until the 1960s, there was no road connecting the Petén to the rest of the country to which it belonged, and its residents were more oriented toward their Mexican and Belizean neighbors downriver than their compatriots to the south.[1] With so much natural wealth in such easy reach of foreigners (and so inaccessible to its sovereign owners), Guatemalan nationalists long feared, with good reason, that their country's natural patrimony in the Petén would be ransacked by outsiders.

The anthropologist Norman Schwartz has described the Petén as the "quintessential frontier"—an isolated zone rich in natural resources and lacking in the rule of law. After the arrival of Europeans, disease wiped out more than 90 percent of the indigenous population, whose carefully tended woodlands went feral, eventually becoming mature rainforest. Loggers arrived in the eighteenth century, attracted by the unusually dense stands of mahogany. In the mid-nineteenth century, mahogany logging spiked, along with the tapping of *chicle*, the original base for chewing gum.

A hyper-masculine and violent frontier culture emerged among the woodcutters and resin tappers, popularized in novels by some of Guatemala's most influential authors. The Petén's reputation for wildness, both in the ecological and social realms, is exaggerated and misleading, but the myths of a primeval lost world portrayed in novels, explorers' travelogues, and even scientific literature are resilient. Frontier myths have driven state policy and popular conceptions about the place since at least as early as the nineteenth century, and still do today.[2]

The first serious attempt to incorporate the Petén into the Guatemalan nation came in the 1940s, when the country's first democratic government tried to escape entrenched colonial legacies by creating a new society of yeoman farmers and lumberjacks in the "empty" northern lowlands. Centered in a combined army base, planned city, and research station in Poptún, in the southeastern corner of the Petén, it was the most expensive of any government program up to that time. It was soon abandoned when a radical turn in the democratic coalition discarded frontier colonization for land reform in the heartland, but it set an important precedent for the authoritarian regimes that emerged as a reaction to that land reform: the colonization project at Poptún charged the army with leading the project and instituted martial law across the Petén to ensure an orderly colonization and development process. Many of the commanding officers who led the more ambitious development program in the 1960s, under right-wing authoritarian governments, had earned their stripes at Poptún in the 1940s as technocratic junior officers of a (moderate) leftist revolution. After the colony was shut down, the army remained, quietly maintaining the base facilities. When the military governments renewed the march north in the 1960s, the Poptún base—once an icon of revolutionary democratic aspirations—became the foundation of authoritarian military power in the Petén.

An emporium of resources: developmentalist counterinsurgency in the 1960s

In the wake of the Cuban Revolution in 1959, security analysts in the United States worried that Guatemala would be the next domino to fall. Since a CIA-sponsored mercenary army had deposed the democratic socialist government in 1954, the country had been plagued by chronic instability, corruption, economic recession, and periodic food shortages. Fearing the imminent rise of a Central American Fidel Castro, the Guatemalan Ministry of Defense, with stepped up US support, began professionalizing and centralizing its internal security services. After months of quietly implanting itself into virtually every branch of the national state, in 1963 the army—as an institution, rather than an individual—ousted the sitting strongman and claimed the role of head of state for itself. The high command set out to permanently institutionalize this arrangement by incorporating reactionary oligarchs and centrist civilian reformers into an ambitious program of military-led economic development. Economic growth, combined with surgically targeted "good works" and repression, was intended to win over the hearts, minds, and bellies of the peasantry, while also eliminating leftist alternatives and disarming the extreme right. At the outset, the military government's flagship project legitimizing its claim to power was the "liberation," as it called it, of the idle and vulnerable forests in the Petén, turning them into an engine of a new industrial economy. The application of science and technology, under the disciplined management of the armed forces, would transform the wild rainforests into rationalized landscapes of industrial timber farms. "Rational planning" of the "emporium of resources" in the Petén, the commanding officers declared, would "give a new structure to the country"—thereby inoculating the peasantry from communism by radically modernizing the economy.

To lead the mission in the Petén, the military revived a mothballed agency called the National Company for the Promotion and Development of the Petén (known by the Spanish acronym, FYDEP) and assigned one of the junior officers from the Poptún colony as its director. FYDEP was a strange bureaucratic amalgam, blending the administrative responsibilities of conservation, research, infrastructure construction, policing, and border security under a single military command that reported directly—and only—to the president. As its name suggested, FYDEP was also a profit-seeking enterprise, selling timber, chicle, and other forest commodities to finance its operations and using surpluses to fund infrastructure and social welfare programs like subsidized housing back in Guatemala City. Its autonomous character made FYDEP a "state within a state," with sovereign and exclusive authority over all people and things in the Petén. Conceived in the 1950s, FYDEP had been modeled on the TVA, that prototypical semi-autonomous, technocratic regional development agency tasked with transforming a neglected backwater into an industrial powerhouse. At its founding in 1958, the reigning dictator had found another use for it: buying the loyalty of a powerful oligarchic family by handing it over to their scion as his own private duchy. When he showed signs of using his agency to build a base of political and military support rivaling his boss, the dictator sacked him and effectively shuttered the company. The military government revived FYDEP and convinced the US Military Assistance Program, the UN Food and Agriculture Organization, and the UN Special Fund to provide funds, equipment, and technical assistance.

FYDEP activities were divided between the objectives of security and rationalizing the wooded landscape. To secure the province from smugglers and subversives, it reinforced and modernized border crossings, increased patrols along the rivers and mule trails that criss-crossed the area, and installed a security "nerve center" at El Ceibo, where the San Pedro river flowed into Mexico. A proprietary electronic communications network allowed officers to coordinate far-flung and isolated patrols in the jungle.

A new headquarters in Santa Elena, in the geographic middle of the province, centralized FYDEP's operations on an air base that permitted the rapid aerial deployment of troops anywhere in the department on minutes' notice. FYDEP's most expensive and technically challenging task was cutting roads through the forest to knit the region into a coherent whole and connect it with the cities and shipping ports to the south. A dedicated Engineering Battalion, trained at Fort Gulick in the US Panama Canal Zone, built roads as well as the "environmental infrastructure" to "sanitize" the tropical lowland environment that was as rich in endemic diseases as natural resources. Workers moved enormous quantities of earth and rock to drain swamps and raise roadbeds. Everywhere that soils did not favor the formation of swamps, water percolated quickly through the karstic limestone bedrock, doing little to filter contaminants. Well water was a common source of communicable illness in the Petén, forcing FYDEP in many places to truck it in and store it in cisterns. Villagers were relocated into "hygienic" planned towns (which were more easily surveilled than their previous communities), where army doctors conducted regular "medical workdays" to treat the sick and inject vaccinations. With transportation and sanitation infrastructure in place, FYDEP could focus on the forests themselves, converting them from ecologically diverse and opaque environments to radically simplified industrial timber plantations.

When soldiers arrived to make a "promised land" of the Petén, none of them knew much about the forests they found themselves in. No one else did either: as of the early 1960s, only cursory scientific surveys had examined the Petén. For the military government, that knowledge gap posed a national security risk: not only would it hinder economic development, it also increased the likelihood that irreplaceable resources might be destroyed through improper use. The army therefore invested heavily in scientific research to render the Petén landscape legible and transformable to its technicians. As one of FYDEP's commanders explained, "in order to transform [the Petén]," they first had to "get to know it better."[3]

FYDEP partnered with the Food and Agriculture Organization to conduct a five-year, comprehensive ecological study to "serve as the basis of rational planning." The constellation of researchers, labs, training programs, experimental milling facilities, and test plots that came to be collectively known as the FAO-FYDEP study was the first planning project of its kind in Guatemala. Based on its initial findings, FYDEP compiled a land-zoning plan, which set about half of its land aside as forest reserve. This included everything north of parallel 17°10', forming the largest contiguous protected area on the American Isthmus.[4] Other reserves were demarcated around the upper branches of the Machaquilá, Pasión, and Usumacinta Rivers, where the densest stands of mature mahogany remained. FYDEP zoned small areas around existing settlements for agricultural development, plus some ranching in the savannah belt between Flores and Poptún, with restrictions on cutting any valuable trees for planting. A special team of scientists set up a "demonstration management area" to train personnel in the latest best practices of scientific forestry, including processing at a large milling facility "in the middle of the jungle," as its supervisor boasted.[5] The demonstration management area, which encompassed a fifth of the Petén's woodlands, overlapped one of the agricultural development zones, further limiting the land permitted for cropping. The scientists in charge of the FAO-FYDEP study cited elaborate statistical and experimental data to confidently predict that timber from the Petén would generate $300 million annually once the forests had been "improved," making it the country's most valuable export, and the wood-based industry the cornerstone of a new society.[6]

FYDEP purchased majority stakes in the three private logging companies working in the Petén, effectively absorbing them as subsidiaries. Each company received cutting rights to a designated bloc of land, subject to the oversight of FYDEP's forest wardens (see Figure 3.2). FYDEP supplied funds and equipment to upgrade the companies' operations to increase the efficiency of harvesting, milling, and shipping. In the northeast, two companies cut, milled, and shipped mahogany and cedar through Belize. In the

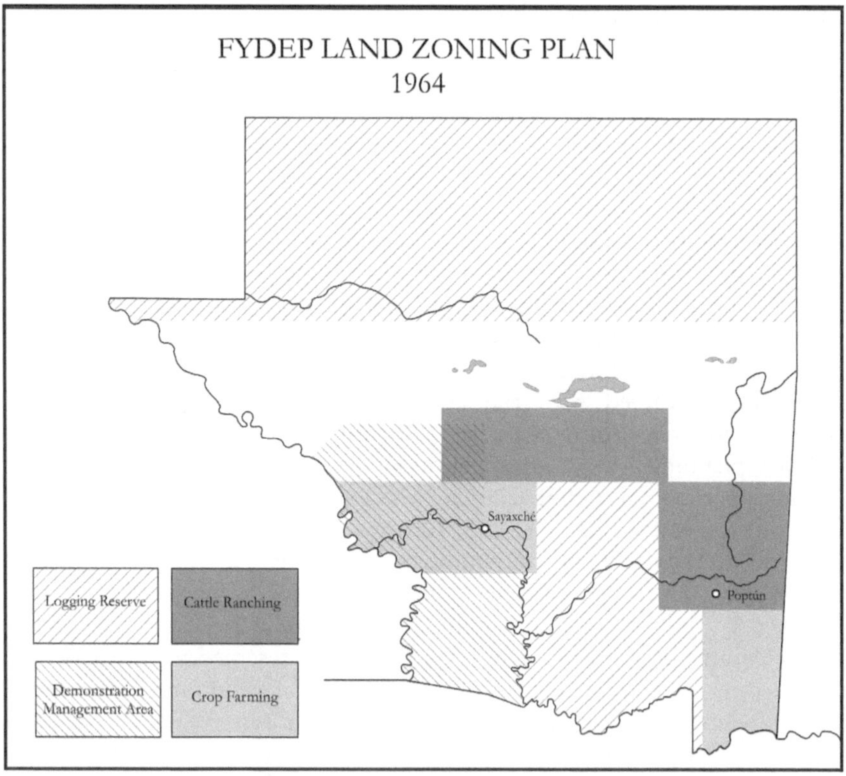

Figure 3.2 FYDEP Land Zoning Plan 1964.

northwest, another company had a concession along the San Pedro River which, due to the swampiness and lack of roads in the area, was granted special permission to export the wood as unprocessed logs floated downriver to Mexico. FYDEP's most productive logging operation, called the Gutiérrez Brothers Lumber Company, was different from the others. In the 1950s, Gutiérrez Brothers (owned by one of the wealthiest families in the country) had processed mahogany, cedar, and other hardwoods into fine furniture, but had shut the company down after just a few years rather than pay the requisite taxes and fees stipulated by their contract. The company's assets were seized and placed under the management of a state auditor, who had increased production and profits by investing in new equipment. FYDEP allowed Gutiérrez Brothers to run more or less independently so long as it paid its taxes.[7]

Although simplifying diverse rainforest ecosystems and logging extensive areas of mature trees might not look like conservation from the perspective of the twenty-first century, it is important to keep in mind that the soldiers and scientists working for FYDEP understood their work to transform the Petén's jungles as an act of conservation. The ideas about tropical nature that informed its policies came from the economistic forestry models of the early twentieth century that were only just beginning to absorb ecosystemic concepts that are familiar to us now. Studies like the FAO-FYDEP project contributed to that scientific evolution, but they had little effect on the Guatemalan army's actions in the 1960s. FYDEP hoped to maximize the efficiency of timber production from one or two species, and acted accordingly to "defend the forest" from those who would waste or despoil its wood stocks. Almost invariably, Guatemala's military governments defined the enemy of the forest as peasants—particularly the indigenous Maya. A FYDEP commander put it bluntly to an audience of ministers and fellow officers: "no matter how much sympathy we may have for the Indian problem, they are not the human contingent that the Petén needs to progress."[8] Officers limited migration to the Petén, allowing only those they deemed rational users. Forest managers and scientists embodied FYDEP's rational ideals, while subsistence farmers were seen as inherently irrational and therefore a threat.

* * *

The anti-communist coalition that supported the military government was far from united in worldview or policy goals, and its internal divisions manifested in the Petén, as well as Guatemala City. Various groups wanted the Petén immediately opened to mass colonization, not locked up in forest reserves. Each faction had its own reasons for wanting to advance the agricultural frontier, and for opposing FYDEP's forest reserves: the oligarchy, whose wealth was tied to large landholdings, believed that colonization would defuse the pressure for land reform; bankers and oligarchs saw an opportunity to speculate in real estate and accumulate more lands; Christian Democrats saw providing land to destitute peasants as a moral obligation, but opposed systematic land reform; foreign agribusinesses lobbied for access to new lands for plantations of palm oil, sugarcane, and cotton; and politically centrist technocrats associated with the first colonization scheme in the 1940s argued that FYDEP could integrate small farmers with forestry, if agronomists were given the authority and resources to supervise the settlers.

The military government convinced its constituents to toe the line on colonization for the sake of public order—until 1964, when Mexico announced plans to dam the Usumacinta river just outside of Guatemalan territory. The resulting reservoir would have flooded thousands of square kilometers in the Petén, which the army interpreted as a violation of Guatemalan sovereignty. In response, the high command ordered FYDEP (against its director's protestations) to settle tens of thousands of peasants along the affected riverbanks, creating a "human wall" to block the dam's construction.[9] Once the gates of FYDEP's sylvan fortress had been opened, the oligarchs joined with the other pro-colonization groups to badger the National Assembly to declare the settlement of the Petén a matter of "public utility and national urgency."[10] The Samayoa family, one of the largest landowners among the oligarchs, somehow persuaded the high command to name one of their own as chief of colonization at FYDEP. FYDEP's unwilling acceptance of an oligarch being granted personal control over a large segment of its domain is important for the analysis here. It indicated first of all, that national security (confronting the Mexican dam) quickly trumped the army's most important domestic policy, and second, that the army included the stability of its political coalition in its security calculations. Reversing its earlier statements on colonization being "demagogical" and irrational, the army declared that "ecological conditions" along the rivers were ideal for small farmers and settled 15,000 in just a couple of years.[11] The gambit to block the dam worked, but at the cost of FYDEP's autonomy and its conservation mission.

FYDEP's nominal conservation and development goals continued to unravel under multiple centripetal forces. By 1965, the oligarchs were openly demanding a heavier hand against guerrillas, distancing themselves from the developmentalist approach favored by the military and organizing paramilitary death squads. Centrists and "apolitical" technocrats kept pushing for colonization as a necessary and just alternative to land reform, and were joined by some military officers who believed such a policy would help tamp down unrest in the countryside. And a new generation of army officers discovered that control over frontier territories rich in natural resources could be used to accumulate personal fortunes rivaling those of the oligarchs. The end of the 1960s saw FYDEP institutionally weakened at the moment that the roads it built opened the Petén forests to spontaneous colonists and land speculators, who cleared land for farming and cattle ranches. Tens of thousands of mature trees had been cut by FYDEP's subsidiaries, representing hundreds of millions of dollars worth of wood, but little discernible "development" was apparent. Dozens of less valuable trees fell for each mahogany log harvested. For all its investments in scientific forestry and the discipline of military oversight, the army had paved the way for deforestation through the very institution that was designed to prevent it.

Scorched earth and fortress conservation in the 1980s

By the 1980s, Guatemala's internal conflict had evolved into an entirely different kind of war. The insurgency had grown from a small group of disaffected officers, confined to the eastern mountains, to a nationwide popular movement that controlled substantial chunks of countryside, particularly in the indigenous Maya villages of the western highlands, just south of the Petén. In 1979, the Sandinistas overthrew the Somoza dictatorship in Nicaragua, raising the prospects of leftist guerrillas across the region. The Guatemalan army responded with increasingly arbitrary and massive violence, working hand in glove with the oligarchs' death squads. Their "100% brute force" method of counterinsurgency culminated in a scorched earth campaign that made no distinction between civilians and guerrillas in "kill zones," which encompassed large swaths of the country. From late 1981 to early 1983, the army killed perhaps 150,000 civilians and sent 1.2 million more fleeing for their lives. One general compared it to "draining the sea of peasants" in which the guerrillas "swam."[12] The lowland forests of the Petén were remote from the highland killing fields, and therefore assumed by many observers to have been peripheral to the counterinsurgency. On the contrary, the army was using conservation in the Petén as an innovative weapon of war. Recognizing this recent past unmasks the uncomfortable legacies of state terror in the conservation landscapes of today, which many conservationists would prefer to ignore.

While burning its way across the indigenous sierra, the army used its logging reserves in the Petén as a green wall to contain the fallout. Many villagers, as predicted by the army, tried to flee to safety in the adjacent lowland jungles. To help channel the refugee flow into prison camps and across the Mexican border, the army deployed an extra 900 special forces soldiers in southern Petén. The soldiers blocked mule trails and navigable rivers, forcing unprepared refugees into the tractless woods where many died of starvation, snakebites, disease, and exposure. But the army could not control everything on the ground. Remarkably, tens of thousands of refugees managed to evade army patrols and survive the hostile environment, eventually setting up autonomous "popular communities in resistance" in the jungle.[13] These autonomous communities were skilled at attracting international human rights networks to their cause, raising their profile and making them "costlier" targets. The army could not simply eradicate them and claim that nothing had happened or that guerrillas and criminals were somehow responsible (as it often did to obscure its own responsibility). The successful autonomous communities were exceptional, however. Most refugees were not so lucky.[14]

By 1983, the army had achieved a kind of pyrrhic victory, leaving it in control over more people and territory, but "bankrupt in every conceivable way."[15] The indiscriminate killing of civilians had damaged strategic alliances, and the international outcry had made the military an international pariah—even its closest ally, United States, had suspended its support (although Israel and Argentina stepped into its shoes).[16] For all the bloodshed, it soon became clear that the guerrillas had not been defeated. Without a civilian support base, the insurgents no longer posed an existential threat, but combatants retreated and quickly regrouped in the jungle, as they had before.[17] The army could not afford, financially or politically, to continue pursuing its scorched earth offensive on the same scale. Even if it could, some officers (and their foreign advisers) believed that the chaos they unleashed was itself a security risk, generating sympathy for the insurgents while alienating allies. Two successive coups, in 1982 and 1983, attempted to correct this course by pushing out the most hardline and destabilizing elements from the military government. The US Reagan Administration supported these changes of the guard, giving diplomatic cover to the new face of the Guatemalan military state. The massacres continued, but in a more deliberate and focused manner. Guatemala's generals coupled their new military strategy with a program to restore their international standing by formally handing over power to elected civilians. The general in charge of this "transition to democracy" made clear that it was not the end of the counterinsurgency, but rather a continuation of it. In 1984, an elected National Assembly convened to write a new constitution, the following year a civilian won a presidential election, and the United States resumed direct military aid to the "democratic" government. Death squad killings coordinated by the high command proceeded as before.[18]

A large system of protected areas in the Petén were integral to the democratic transition, crowned by the 1.7 million hectare Maya Biosphere Reserve (MBR). The MBR was created from FYDEP's logging reserve north of parallel 17, with a "buffer zone" added along its southern edge. The final borders of the MBR were moved to either swallow up villages unfriendly to the army or leave sympathizers alone.[19] The army liquidated FYDEP and transferred its logging reserves, equipment, and personnel to new civilian administrative agencies, but the army retained its role as law enforcer. Peasants fleeing into the woods were discursively rendered as "invaders" and "environmental criminals," rather than refugees or citizens who could assert rights to public lands. With few exceptions, peasant farmers could not legally claim land in the Petén as part of any peace and reconciliation process, because those lands were protected as woodland. Early in the peace negotiations, before the biosphere proposal had gained traction, the army had agreed to resettle some refugee communities in northwestern Petén. Those army-sanctioned returnees were later evicted by the army for having "invaded" the biosphere reserve.[20] Autonomous communities, which had survived the army's aerial bombings and spraying of toxic herbicides, were dismantled in the postwar for having illegally squatted in protected forest. Some of the first agreements in the peace negotiations stipulated the demobilization of the army's special forces, which were responsible for countless atrocities, but the army found new uses for them in the Petén, where several units became forest wardens. In their new uniforms, they continued to capture and kill peasants (as well as traffic illicit timber and drugs), but with a good deal less international outcry.[21] Conservationists were shocked by the violence, but many justified it on the grounds that the Petén's untrammeled wilderness—the patrimony of *all* humankind—had to be defended.[22]

A host of international actors played vital roles in transitioning the Petén from military to civilian administration. The concept of the biosphere reserve originated in the 1970s in the context of the early global environmental movement, which hoped to supplant economistic models of conservation with ecologically and socially informed ones. Conservation International, a large international NGO based in Washington, DC, first proposed the Maya Biosphere Reserve in the mid-1980s and arranged with multilateral lenders for a "debt-for-nature swap" to incentivize the civilian government to accept their plan. The local directors of Conservation International perceived a brief window of opportunity to act before either corrupt officials grabbed all the land in the Petén for themselves or the war intensified again, making the area too dangerous to work in.[23] The Nature Conservancy and the World Wildlife Fund joined Conservation International to put up $15 million to match the government's $7 million contribution. The US Agency for International Development offered a $20 million loan with generous terms to keep the project on pace when *both* corruption and political violence threatened to torpedo the plan.[24] Although the leaders at Conservation International felt they had achieved their goal only at the eleventh hour, the timing was just right for their camouflaged allies.

* * *

The army's embrace of conservation did not mean that it had internalized the environmental ideals of its NGO partners. Forests could be a green wall for the army, but they provided cover to insurgents. The army burned down the woods around the villages it targeted, and did the same to increase visibility around its own encampments.[25] In the Petén and elsewhere, army planes dropped napalm and herbicides over remote autonomous communities, where it claimed guerrillas were cultivating marijuana.[26] Special forces soldiers hunted wild game and captured endangered birds in their leisure time.[27] Between 1979 and 1987, FYDEP reported logging 100,000 mature hardwood trees—18,000 of which were felled in 1987 alone.[28] Spontaneous colonization, accelerated by counterinsurgent violence, had driven a tenfold population increase in the Petén from the 1960s to the 1980s.[29] The influx of people dramatically increased the human footprint on the forests, particularly along the roads built by FYDEP across the southern Petén for its logging operations, where colonists cleared vegetation to plant subsistence and cash crops.

The natural resources under army control in the Petén were lucrative for the officers who guarded them. Beginning in the 1970s, multinational oil companies, with high-ranking army officers as investor–partners, had been extracting small, but increasing, amounts of crude from the northwestern corner of the Petén. Officers were quick to blame spills and fires on insurgent sabotage (which did happen), but leaks and accidental fires happened frequently without any outside interference.[30] Officers expanded their own estates into adjacent protected areas by dangling the promise of free land to peasants, who cleared and prepared it for planting. They would then be summarily evicted as environmental criminals, while officers began planting palm oil and other export commodities. Conservationists tended to blame the peasants and defer to the officers who had taken final possession of the land. In place of plantations, some officers established "narco ranches," which are exactly what the name implies, and in others they set up ecotourism venues, earning the enthusiastic support of prominent conservationists. It was an open secret that FYDEP's top officers extracted more trees than they reported, selling them on the black market and pocketing the profits.[31] The scale of deforestation caused by unmitigated logging is routinely ignored by conservationists who blame peasant "slash and burn" farming and population growth for the retreat of the forest. All of these outcomes, especially taken together, appear to directly contradict the goals of conservation, but the actions taken by the army reflected a particular set of values about nature and who was permitted to use it.

The establishment of the Maya Biosphere Reserve coincided with the last phase of the war, as peace negotiations between the guerrillas and the army gained momentum (see Figure 3.3). The peace accords, signed in 1996, included a framework to address the inequalities and injustices that had fueled the insurgency, but they did not call for land reform and the large conservation areas in the Petén undermined their aspirations for justice. The militarized conservation landscapes did not necessarily lead to effective conservation outcomes, as defined by international practitioners, but they did enrich the war's victors. Today they are some of the most violent places in the Guatemalan countryside, and even the most ardent conservationists admit that deforestation is accelerating in spite of all the environmental laws and institutions. Authoritarian conservation in the Petén, in its many forms, was an effective weapon of counterinsurgency, but only selectively so as a tool to combat environmental degradation.

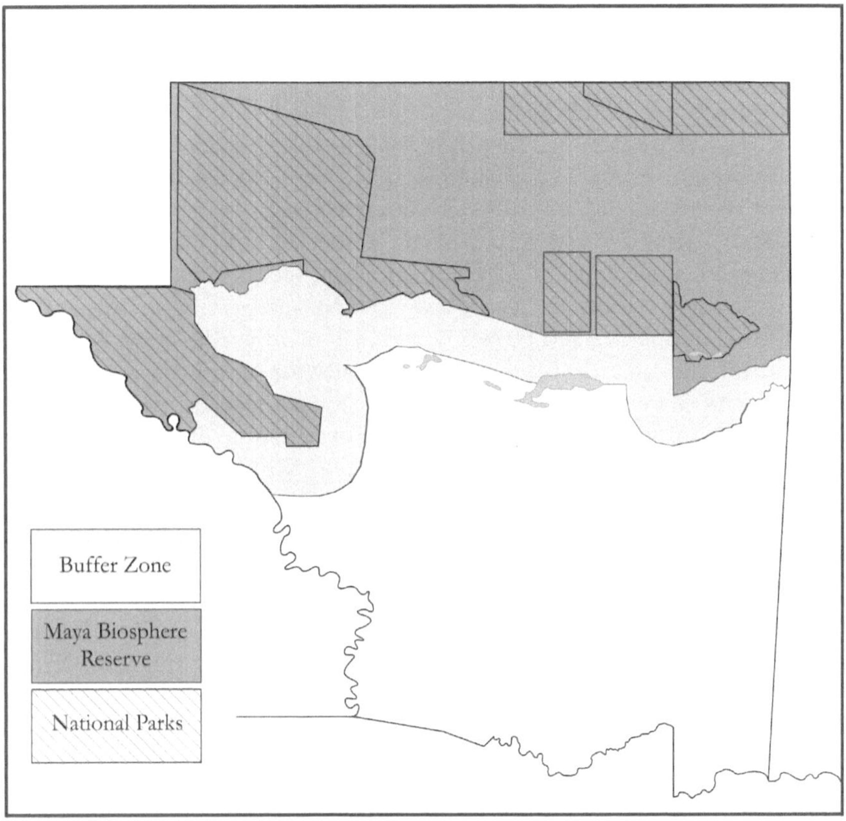

Figure 3.3 The Maya Biosphere Reserve.

Shrouded legacies

In the summer of 2017, a thousand Guatemalan soldiers and police descended on the village of Laguna Larga, in the far north of the Maya Biosphere Reserve. The villagers, warned of the approaching forces, fled with what they could carry on their backs, taking refuge at the border where they believed the army would not risk a diplomatic spat by attacking them. The villagers had been living in Laguna Larga since the 1990s, and had negotiated with conservation administrators and various state authorities to receive a community forest concession, which they would protect and maintain in exchange for the right to live in the reserve. But in 2017, the Guatemalan chapter of the Wildlife Conservation Society (best known as the organization that runs New York's Bronx Zoo) filed suit in Guatemala City, alleging that the village represented an "existential threat" to endangered wildlife and archeological artifacts in the area. Acting on a civilian judge's orders, the soldiers and police burned the village's homes and crops to the ground, leaving the town's chapel intact as a permanent security outpost. The Wildlife Conservation Society praised what it called the triumph of the rule of law, but few in Guatemala failed to notice the echoes of genocide.[32]

For most of the last half century (1968–2018), conservation policies in the rainforests of northern Guatemala have been authoritarian, but they have not been ideologically consistent nor effective in the terms understood by most North American or European conservationists today. Whether justified by the principles of developmentalism, scorched earth, or a return to democratic governance, it was bound and subservient to the army's imperatives of political and economic control over the peasantry. Rival factions within the military state battled constantly for supremacy so that their material and ideological interests might become those of the state itself, resulting in unstable policies and the reappropriation of institutions for purposes other than those originally intended. Military officers incorporated the latest scientific discourses into their conservationist programs, altering the form of military administration of woodlands, but they readily discarded scientific counsel when it conflicted with the social hierarchies they enforced. Liberal democracies like the United States, West Germany, and Israel provided the funds, arms, scientific advisers, and institutional models for authoritarian conservation in Guatemala, incriminating "free" societies in the maintenance of despotic regimes of environmental regulation—*and* ecological degradation.

The dark origins of the Maya Biosphere Reserve are by no means unique. Similar stories are symptomatic of large conservation projects across the postcolonial world, where indigenous populations have been displaced and excluded from their homelands by colonial governments, which then walled them off as game reserves, timber plantations, or for their aesthetic beauty. And as in Guatemala, conservation has been used all over the world as a coercive apparatus to integrate rebellious hinterlands into the state and global markets. C. S. Lewis's memorable statement, that "man's power over nature" is in reality nothing more than "one man's power over another, with nature as its instrument," applies to all conservation policies, in any society, free or unfree. In some places this power is just more visible.

Notes

1 Norman Schwartz, *Forest Society: A Social History of Petén, Guatemala* (Philadelphia, PA: University of Pennsylvania Press, 1990).
2 See Schwartz, *Forest Society*, chapter 4.
3 Remarks delivered by Colonel Oliverio Casasola at the opening of an industrial sawmill, quoted in both *El Imparcial* and *Prensa Libre*, June 8, 1968.
4 Schwartz, *Forest Society*, 252–253. See also "Informe Final," *Estudio de Preinversión sobre el Desarrollo Forestal* (Rome: FAO, 1970).
5 *El Imparcial*, December 16, 1966.
6 Equivalent to $2 billion in 2017 dollars.

7 See *El Imparcial*, March 23, 1954; February 17, 1956; March 10, 1959; June 13 and July 21, 1962; September 22, 1964; September 4 and 5, 1966; and May 25, 1967.

8 Quote from Melville, *Politics of Land Ownership*, 223–224. Originally from *Grandezas y Miserias*, given in 1967.

9 Peralta's plan was not to use Guatemalan bodies as a literal wall against rising water, but rather to increase the indemnification costs to Mexico, as required by the international treaty for damage to Guatemalan territory, to a prohibitive level. Peralta ordered the construction of the "human wall" through Decree 266, in August 1964. Decree 266 can be found in the legal records of the Archivo General de Centro America's Hemeroteca. Its passage was accompanied by extensive propaganda in contemporary newspapers. Peralta set a ten-year deadline from June 1965 to complete the project, via Decree 354, which appears in partial translation in Melville, *The Politics of Land Ownership*, 161–162.

10 Melville, *The Politics of Land Ownership*, 157–158. See also the series "Problemas y Desarrollo" on the Petén in *El Imparcial*, 7, September 21–23, November 3, 8, 9, and December 5, 1964.

11 This settlement process is covered in extended detail in Melville, *The Politics of Land Ownership*, and Millet, "The Agricultural Colonization of the West Central Petén, Guatemala: A Case Study of Frontier Settlement by Cooperatives." I cover FYDEP's tensions with the high command in chapter four of my dissertation, "Environmentalists with Guns: Conservation, Revolution, and Counterinsurgency in the Petén, Guatemala, 1944–1996," New York University, 2018.

12 See Schirmer, *The Guatemala Military Project*, chapter 2.

13 See *Comisión para el Esclarecimiento Histórico* (New York: United Nations, 1994); "Plan Victoria 82," in the CIRMA Historical Archive, Antigua, Guatemala; and the Defense Intelligence Agency, "info report," August 10, 1982, available at the Digital National Security Archive, https://search.proquest.com/docview/1679120857.

14 "'We are Civilians': Communities of Population of Resistance of the Sierra," in *The Guatemala Reader*, 427–430. For the communities in the Petén, see "Nuestra lucha es semilla del futuro," CIRMA Historical Archive.

15 Carlotta McAllister and Diane Nelson, "Introduction" to *War by Other Means: Aftermath in Post-Genocide Guatemala* (Durham, NC: Duke University Press, 2013), 34.

16 It should be noted that the US State Department wanted to provide military aid, but felt that it could not, given the reputation of the Guatemalan military. See "We Cannot Confirm nor Deny," in *The Guatemala Reader*, 378–385.

17 Susanne Jonas, *The Battle for Guatemala*, 148–150.

18 General Héctor Gramajo called the transition to democracy an inversion of the famous aphorism of the Prussian strategist Carl von Clausewitz. Clausewitz had called war a continuation of politics "by other means." Guatemala's generals were using politics to continue the counterinsurgent war. See "Inverting Clausewitz," in *The Guatemala Reader*, 417–420.

19 Notably by adding in the Sierra del Lacandón National Park, along the Usumacinta River, which corresponded to the territory most thoroughly controlled by the Rebel Armed Forces (FAR) guerrillas in the Petén.

20 Pierre van der Vaeren, *Perdidos en la Selva* (Amsterdam: Thela Publisher, 2000).

21 See Jennifer Schirmer's and Edgar Gutiérrez's chapters in R. Sieder (ed.), *Guatemala After the Peace Accords* (London: University of London, 1998).

22 See Megan Ybarra, "Living on Scorched Earth: The Political Ecology of Land Ownership in Guatemala's Northern Lowlands," PhD Dissertation, University of California, Berkeley, 2010; and Juanita Sundberg, "NGO Landscapes in the Maya Biosphere Reserve, Guatemala," *The Geographical Review* 88, no. 3 (1999):

388–412; and Liza Grandia, *Enclosed: Conservation, Cattle, and Commerce among the Q'eqchi' Maya Lowlanders* (Seattle, WA: University of Washington Press, 2012).

23 Pereira, *Unfinished Conquest*, 224–225.

24 It is rumored that the CIA played a role in arranging the USAID loan, which was approved and instituted at unusual speed for a project of its kind. Personal communication, Ronald Nigh, 2015.

25 Jonas, *The Battle for Guatemala*, 149.

26 Fuerzas Armadas Rebeldes, "También sufre el Petén," and the various communiqués issued by the CPR-P in the early 1990s. Both held at the CIRMA Historical Archive.

27 Victor Pereira, *Unfinished Conquest: The Guatemalan Tragedy* (Berkeley, CA: University of California Press, 1993), 239–240.

28 Pereira, *Unfinished Conquest*, 240.

29 Actual population figures are somewhat murkier than this might suggest. But it is unquestionable that migration into the Petén was extensive and prolonged for many years. See Liza Grandia, "Cuantas personas quiere usted que vivan en el Petén?" in *Nuevas Perspectivas de Desarrollo Sostenible en Petén* (Guatemala City: FLACSO, 2000), 137–158.

30 Solano, *Petróleo y Minería en las Entrañas del Poder*. See also Amy Rosenfeld, "Oil Exploration in the Forest," in J. D. Nations (ed.), *Thirteen Ways of Looking at a Tropical Forest: Guatemala's Maya Biosphere Reserve* (Washington, DC: Conservation International, 1999), 68–76.

31 United States Defense Intelligence Agency, "Military Corruption in the Petén," March 1995; and United States Defense Intelligence Agency, "Continued Involvement of Military in Illegal Activities," May 1995. Both available at the Digital National Security Archive's files "Death Squads, Guerrilla War, Covert Ops, and Genocide: Guatemala and the United States, 1954–1999," https://search.proquest.com/dnsa_gu. Accessed: August 26, 2018.

32 Greg Grandin, "A Brutal Expulsion in Guatemala Shows How Neoliberalism Gets Greenwashed," *The Nation*, June 8, 2017.

4 Hydroelectric dams and the rise of environmentalism under dictatorship in Brazil and Paraguay (1950–1990)

The case of Itaipu

Nathalia Capellini Carvalho de Oliveira and Carlos Gomez Florentin

The construction of the Itaipu Dam offers an interesting case to understand how right-wing dictatorships grasped and dealt with the rise of environmentalism in the 1970s. This chapter explores the emergence of both environmental legislation and organized struggles in the context of two South American authoritarian regimes. The desire for a final settlement of still-unsettled borders, the sealing of a new alliance after decades of suspicion and animosity, and the state-developmentalist push of both Brazilian and Paraguayan military governments all contributed to the creation of the Itaipu Dam. That is, political and economic factors were the principal driving motives behind the dam's construction, and environmental concerns were minimal at best. While this is not surprising, the unanticipated impact of global political and intellectual currents created, within the military regimes, legal and discursive frameworks for addressing environmental concerns during dam construction—concerns that both regimes preferred to leave as empty letters. However, the social and political mobilization of affected populations compelled dam builders to take real account of environmental factors, fulfill some environmental laws and regulations, and even create entirely new bureaucratic infrastructures to address them, thereby fundamentally transforming the actual experience of dam construction and operation in ways that the original visionaries of hydroelectric modernity did not intend nor foresee.

The first part of this chapter explores the rise of military regimes and developmentalism in both Paraguay and Brazil and details the long process that led to increasing consensus in both Brazil and Paraguay to opt for hydroelectricity as the main source of energy for the region. The second part addresses the ripple effects of a global framing of the environmental question that emerged in the 1970s and its unfolding in dam construction, especially in the Itaipu case. Finally, the chapter explores the trajectory of an emerging environmental movement and its relationships with decaying dictatorships in the two countries and highlights the lasting legacy of dictatorships in the management of nature.

Military regimes in Brazil and Paraguay: development plans and hydroelectricity expansion

The tense relations between Brazil and Paraguay that had characterized most of their bilateral history permeated by conflict, suspicion, and war, only eased with the rise of military dictatorships in both countries during the second half of the twentieth century. But the alliance between the two governments was really sealed when Paraguay under General Stroessner (1954–1989) and Brazil ruled by a military junta (1964–1985) launched the ambitious Itaipu Dam project on the Paraná River in the 1970s. The biggest dam in the world, until the inauguration of the Three Gorges Dam in 2008 on the Yangtze River in China, the Itaipu Dam has an installed capacity of 14,000 MW. Built between 1974 and 1984, the construction of the dam involved more than 31,000 workers, used 12.57 million square meters of concrete and cost more than US$20 billion. The 1,350 square kilometer reservoir displaced 40,000 people, flooded most of the Avá Guarani indigenous lands, and submerged 780 square kilometers of subtropical forests and agricultural land on the Brazilian side alone.[1] The reservoir also flooded what was once the most powerful waterfall on Earth, the Guayrá/Sete Quedas Falls, located on the border of those two countries, about 170 kilometers up river, triggering an environmental catastrophe that had tremendous consequences.

In the second half of the twentieth century, most South American countries were under dictatorial governments. Paraguay fell under the authoritarian rule first, as it was run by General Alfredo Stroessner between 1954 and 1989. The Brazilian military regime resulted from a civil–military coup d'état in 1964, and it lasted until 1985. Together with authoritarian governments, the "national-developmentalism" strategy was largely applied in the region. This strategy referred to the state's planning of the country's economic growth, participation in key sectors of economy and promotion of industrialization through import substitution. This approach emerges in a postwar context where economic growth had become a key idea to State action, and Latin American countries, excluded from the Marshall Plan, suffered from the abandonment of their industrial park by the lack of exports. Developmentalism was rooted in the idea that industrialization was the easiest way to "burn" the stages and reach development faster.

The first statist to adopt this strategy in the region was Getúlio Vargas in Brazil (1930–1945 and 1950–1954). But soon the idea spread throughout the region, reaching countries like Argentina, Mexico, Chile, Colombia, and Uruguay, especially after the 1950s, when the theories developed by the United Nations Economic Commission for Latin America and the Caribbean (ECLAC) came about.[2]

Both countries embraced a nationalist-developmentalist strategy during the second half of the twentieth century. According to the nationalist-developmentalist model, the state played a major role as the driver of economic modernization. The state had two major lines of action: protect the internal economy from competition from abroad and nationalize key sectors of the economy previously disregarded by private investors. Stroessner's Paraguay also embraced a national-developmentalist strategy, even though it did not achieve industrialization by import substitution, as neighboring countries did, due to a lack of significant internal markets required to sustain the process. The country was among the poorer of the region, lacked substantial industrial development, and had a very small, mostly rural, population in the 1950s. In Brazil, the repressive military government that centralized powers in the executive, applied developmentalism by the book, seeking to modernize the country. As Bresser-Pereira points out, however, state intervention in this case had nothing to do with collectivization of means of production as forwarded by the socialist model, but rather with investing in strategic sectors that were "abandoned" by the private sector to foster capitalist economic growth.[3]

As a result, Brazil experienced a period of exceptional growth between 1968 and 1973 known as the "Brazilian miracle," where GDP grew at around 10 percent a year. The consumption of electric energy increased significantly from the 1960s and even more so following the economic "miracle." The spiking consumption, coupled with the industrialization policy, led to a series of projects to increase energy generation in Brazil. The same happened in Paraguay where the national government built the Acaray I and II dams on the Acaray River, near the frontier region of the two countries, with Brazilian economic aid.[4] During the military dictatorship the hydroelectric installed capacity was multiplied by a factor of seven, going from 4,894 MW in 1964 to 37,437 MW in 1985. This development was unique considering the lack of alternative sources of energy in the country, a situation different from the cases of petroleum-rich countries such as Venezuela and Mexico.[5] Brazilian hydroelectric generation in 1985 was equivalent to 89 percent of the total hydroelectric generation from the whole Central and South American region where, for instance, Argentina generated only 6,000 MW and Venezuela 8,000 MW. Compared to other regions, Brazil hydropower generation is also exceptional—in 1985 France generated 18,000 MW, China 25,000 MW, and India 12,000 MW.[6]

During the "miracle," and especially under the presidency of Emílio Medici (1969–1974), the regime used economic growth to legitimize its political rule. In the media, slogans such as "Powerful Brazil" (Brasil Potência) and "Go Forward Brazil" (Para frente Brasil) were repeated in a nationalistic frenzy. The military political project was based on the idea that a huge territory, with a large population and plenty of natural resources, under the command of the military, could place Brazil among the great powers of the world. The government was convinced that the economy would continue to grow steadily in the years to come. As Bôa Nova points out, this thought served as the basis for the public sector, "imbued with the mystique of a powerful Brazil," to embark on enormous projects, in size and cost, with profits that were sometimes doubtful for the nation and catastrophic for the environment, such as the Transamazônia highway, the Rio-Nitéroi Bridge, the Angra nuclear power plant and several hydroelectric dams, such as Itaipu.[7]

After the oil crisis of 1973, the hydroelectric option became more relevant as a way of increasing the electric supply, boosting industrialization while fostering independence from foreign fossil fuel. The II National Development Program (II PND) for 1975–1979 highlighted the development of a basic industry to replace imports in this area and improve exports of manufactured products. The focus was on steel, metallurgy and petrochemical industries, with state-owned enterprises playing the leading role in this policy. Reaching newer sources of energy meant reaching new territories at the same time. The government targeted new territories such as the Amazon Region and the Upper Paraná Region in its quest for exploiting natural resources.[8]

The increasing energy demand for its industrial plans forced the Brazilian government to take measures. As the oil crisis developed in 1973, Paraguay and Brazil signed the Itaipu Treaty that led to the construction of the Dam in 1974. If Paraguay's economy was also growing, the Itaipu project responded more to a cash flow necessity than a need to fuel an industrial project. A concerted developmentalist vision emerged, no less under the two military dictatorships, launching the greatest megaproject in the hemisphere.

Dreams of development and hydroelectric power in Brazil and Paraguay

The lands that currently belong to Paraguay remained a contested frontier during colonial times as colonial settlers and Indians defended Spanish territory from *bandeirantes*'[9] raids seeking Indian slaves and where Portuguese diplomats sought to carve out territory for their Southern Cone empire. The situation only worsened in the 1800s when Paraguay became independent from the Spanish Empire and Brazil became the home of an exiled Portuguese Crown. While the early dictatorship of Doctor Gaspar Rodriguez de Francia (1811–1840) and the authoritarian rule of Carlos Antonio López (1842–1862) maintained distant, yet civil, relations with Brazil, overt conflict broke out in 1864 under the leadership of Francisco Solano López, leading to the disastrous Triple Alliance War (1864–1870) that pitched Brazil, Argentina, and Uruguay against Paraguay. During the War, Paraguay lost between 60 and 69 percent of its population and 40 percent of its territory.[10] After the war, in 1872, Paraguay and Brazil sorted out frontier issues. But the Loizaga-Cotegipe Peace Treaty left unresolved the ownership of the mighty falls on the Paraná River, known as Salto do Guayrá in Paraguay and Sete Quedas in Brazil. At the time, the issue was dismissed as minor considering that the Paraná River was unnavigable above that region.

But after the rise of hydroelectric dams around the world in the late 1800s and early 1900s, the falls came to represent the promise of hydroelectricity for the entire region. Later negotiations and border demarcations in the 1920s and 1930s thus did not end the border dispute, leaving contested terms for the critical four kilometers along the falls.[11]

The possibility of building a hydroelectric dam on the border between Brazil and Paraguay was an old dream in both countries. In the 1920s, Paraguayan engineer Bruno Mariotti wrote about the potential of the Salto del Guayrá and argued for a shared exploitation of a hydroelectric dam in the zone. His ideas were disseminated among the Paraguayan authorities and laid the basis for building a consensus within the intellectual elite of the country regarding the hydroelectric potential of the country's rivers. At the time, ideas of development centered around energy sources came to the forefront as Brazilian authorities inaugurated several hydroelectric dams in the São Paulo State, such as the Billings dam in the 1920s. News of the dam came to the attention of Paraguayan authorities, among them engineer Albino Mernes from the National Department of Engineering, who shared these ideas with colleagues and the political elite of the time.[12]

On the Brazilian side, hydroelectricity had been developing since the 1910s around urban and industrial centers and next to mining projects. The Water Division of the Ministry of Agriculture of the Brazilian government asked the Office of Inter-American Affairs for a report on the hydroelectric potential of the Sete Quedas Falls as early as 1938. Doctor Antonio Alves de Souza, who wrote the report, considered that the lack of nearby markets for the electricity did not justify undertaking a dam at the time. By 1952, the situation was changing as agriculture progressed in Northern Paraná, Western São Paulo, and Southern Mato Grosso, spurring cities demanding energy.[13]

While engineers in both countries agreed on the need for hydroelectricity, the governments did not agree on who was the owner of the powerful Guayrá/Sete Quedas Falls. Conflicting interpretations went back to the signature of the Loizaga-Cotegipe Treaty in 1872. Successive failed attempts to delimit the frontier between the two countries in 1874, in the 1920s and in the 1930s could not settle the difference as two different cliff lines crossed the Paraná River at the Guayrá/Sete Quedas waterfalls, resulting in irreconcilable positions.

In the 1950s began the surveying of the Alto Paraná borderlands along the upper Paraná River on both margins. While the Brazilians envisioned the exploitation of the Alto Paraná River for electricity in its upper portion around the Paranaiba and Rio Grande rivers, Paraguayans surveyed the potential of the Monday and Acaray rivers at the confluence with the Paraná River. In both cases the increasing demand for energy triggered the decision to invest in hydroelectric projects.[14]

Paraguay in the 1950s remained a sparsely populated, underdeveloped and traditional society. As such, it was an ideal place to attempt development ideas. Soon international aid reached the country to put those ideas into practice. The Paraguayan government was under the still-tenuous early years of General Alfredo Stroessner's leadership, who needed help in consolidating his regime. In this regard, the Brazilian government's aid would come in handy. Brazil had tried to reach out to Paraguay since 1941, when President Getúlio Vargas became the first Brazilian President ever to visit the country. By the 1950s Brazil developed a new "vital interest" on rapprochement with Paraguay after decades of diplomatic mistrust, especially after the catastrophic Triple Alliance War (1864–1870).

The sudden interest in reaching out to Paraguay stemmed from the need to take control of the powerful Guayrá/Sete Quedas Falls and its hydroelectric potential. According to the Argentinean Ambassador in Paraguay at the time, Felipe Yofre, the need for energy triggered the closer links between the two countries. And even though the US Ambassador considered that the potential was there, they all thought no market at the time could possibly consume all the energy that could be created with the construction of the dam.[15]

The studies for the exploration of the Paraná River on the Brazilian side that were made in the previous decade were only presented in 1961 and 1962 during the progressive left-wing presidency of João Goulart. During a meeting between Goulart and Stroessner, the presidents agreed that "no water use in the frontier region would be made without mutual understanding."[16] Immediately after the military coup of 1964, the project was first left aside because of the early national urgencies of building consensus around the authoritarian project. Nevertheless, the energy issue only loomed larger as the industrial needs of Brazilian economy increased. Furthermore, the energy race also involved Argentina, the southern neighbor and also owner of the Paraná River. The regional energy race forced the Brazilian and Paraguayan militaries to reach an agreement that frustrated Argentine intentions to exploit the Paraná River. While the projects down river such as Corpus were left aside, Paraguayan and Brazilian authorities moved forward with the Itaipu Dam upriver. The size of Itaipu, by far the most ambitious dam project of the twentieth century, sent a message that Brazil would not hold back on its developmental goals to keep amicable neighbors' relations with the Argentine government.

Negotiating a project

The Paraguayan government alone lacked the resources to undertake the exploration of energy sources in eastern Paraguay. To do so it needed cooperation from either Brazil or Argentina. Furthermore, Paraguayans needed a shorter path to the Atlantic Ocean to reduce its dependency on Argentina to reach foreign markets. Up until the mid-1900s, the only viable trade route to reach the Atlantic was the Paraguay–Paraná river system. Brazil held the key to an eastern road to the Atlantic shore. As Paraguay sought a way out eastwards, the Brazilian government proved open to cooperate with its landlocked neighbor.

In October 1954, the Paraguayan government made three requests of the Brazilian government: (1) the building of the Coronel Oviedo—Foz de Iguaçu road in Eastern Paraguay; (2) exploratory studies on the hydroelectric potential of the Acaray and Monday rivers; and (3) the return of historical documents taken by Brazilian invaders during the Triple Alliance War (1864–1870). The Brazilian government agreed to the first and delayed the last. The second request, while not granted initially, overlapped with Brazil's plans, which was already on a quest for cheap energy from exploiting the countries' rivers.[17] Continuing the rapprochement between the two countries, the Paraguayan government reached an agreement with the Brazilian one in January 21, 1956 on three critical fronts: free ports, highways, and power development.

Between 1957 and 1965, the two countries sustained cooperative relationships as the Brazilian government built the Friendship Bridge connecting the cities of Ciudad del Este (then called Presidente Stroessner on the Paraguayan shore) and Foz de Iguazu (on the Brazilian shore) across the Paraná River. Most of the projects happened in the region surrounding the mighty Paraná River, but the two governments also cooperated in other areas.[18]

But in early 1962 the major dispute for the possession of the Guayrá/Sete Quedas Falls between the governments of Paraguay and Brazil was already out in the open. The diplomatic rapprochement almost collapsed in 1965 as the Brazilian government manifested its intention of harnessing alone the hydroelectric power of the Guayrá/Sete Quedas Falls. The undecided ownership of those waterfalls was finally settled when the two countries agreed on exploiting the river in partnership through the "Acta de Iguazu," endorsed by the military governments of Brazil and Paraguay in 1966. By a political decision the Guayrá/Sete Quedas Falls became binational, or better said, transnational, as the two countries claimed equal ownership over the energy of the Paraná River.[19]

The developmentalist drive of the two governments proved decisive in the shared decision to harness the hydroelectric potential of the falls. The ideals of "Great Brazil" and the "Second Reconstruction" of Paraguay coincided as the two countries moved forward with the megaproject of the Itaipu Dam.[20] The construction of the dam with Paraguay had a double justification in the minds of the Brazilian political intelligentsia: on the one hand, it allowed them to settle the frontier disagreement for good; on the other, it gave the Brazilian government the cutting edge in the geopolitical race with Argentina to sway landlocked Paraguay into its growing political orbit.[21]

Even though the project involved the flooding of the Guayrá/Sete Quedas Falls, considered then to be one of the most beautiful waterfalls in the world, the decision to move forward with the megaproject was taken in

1973, under military rule in the two countries, without much controversy. The arguments for the decision were fourfold. First, the builders claimed that, since the neighboring Iguaçu Falls were already developed in terms of basic infrastructure and better communication links, there was no point in building an entire touristic infrastructure to exploit the Guayrá/Sete Quedas Falls as well. In a way, this argument exorcized the pangs of consciousness that bothered the builders of the Itaipu Dam regarding the flooding of a natural beauty. A second reason referred explicitly to increasing energy needs in a context when oil prices went up after 1973. A third reason regarded the need to settle delimitation problems between the two countries. Finally, the most important reason was the ability of the project to make nature productive in a way that fitted both dictatorships' ideologies.

José Costa Cavalcanti, one of the main Brazilian architects of the construction of the Itaipu Dam, when asked by a local journalist about the rationale behind the decision to flood the Guayrá/Sete Quedas Falls for the sake of economic development, justified the project this way:

> Sete Quedas is a great work of God but it is producing too little. Barely some basic tourism. Too little compared to Foz de Iguaçu and the waterfalls. Now through man's work, the unproductive work of God will start paying off for the owners of the Sete Quedas that are the Brazilian and Paraguayan governments.[22]

From this perspective, the improvement of untouched nature was necessary to pay off its owners. Moreover, the improvement of nature could not be left to the individual action of nation-states working independently. In this case, a truly transnational institution, neither fully Brazilian nor only Paraguayan, had to take over the construction of the megaproject. The hydroelectric dam was to be managed by a binational state company, the Itaipu Binacional created on April 26, 1973, by the "Treaty of Itaipu." This treaty, signed by Presidents Medici and Stroessner, placed the entity under the responsibility of Administracion Nacional de Electricidad (ANDE) and Eletrobrás, the respective Paraguayan and the Brazilian state-owned electricity companies.

The legacy of Stockholm

At the same time as Itaipu's project was taking shape, the world was experiencing the surge of the idea of a global environment in peril. Even if both countries were living under the yoke of military regimes that carried out an aggressive developmentalist strategy with a productivist approach to nature, the environmental debate was emerging. Despite censorship and repression both dictatorial governments responded ambiguously to that new context. The Brazilian participation in the United Nations Conference on the Human Environment at the Stockholm Conference in June 1972 and its consequences were revealing of the way that a military dictatorship dealt with the new framing of the environmental question. The meeting at Stockholm brought together 12,000 official representatives from 113 nations concerned by the impact of human activity on the environment. Participants signed a twenty-six-principle declaration and agreed that the main threats to the environment were waste discharge, air pollution, and the depletion of the ozone layer. Brazil's participation in the conference was largely portrayed by the media and by analysts as anti-environmentalist, since it put to the forefront its pursuit of economic growth and the country's industrialization. Two ideas summarize what is remembered of the Brazilian participation: first, the then popular idea of "pollution as the worst kind of misery" and second, that environmental issues were seen as a luxury of developed countries that Brazil could not afford.[23]

In fact, the position Brazil defended in this event was consistent with those from other developing countries. The Brazilian diplomats, very outspoken, focused their participation on three issues: they argued against the "zero growth" idea, considered harmful for underdeveloped countries; they implied that environmental concerns could in fact hide imperialist interests against the development of poor countries;[24] and presented the idea that environmental protection depended first and foremost on economic growth.[25] Therefore, for Brazilian representatives, air pollution notwithstanding, the urgency of developing the country could not be trumped by environmental concerns. These arguments prevailed and the final declaration, acknowledged in the fourth issue, was that most of the environmental problems in developing countries such as Brazil were caused by underdevelopment.

Despite the marginal place environmental questions were given in the developmentalist plans of the military regime, it is at this time that environmental legislation emerged in Brazil and began shaping today's institutional framework (even if some previous legislation existed).[26] Between 1965 and 1970, the military regime reformed the forest, hunting, and mining codes, establishing for the first time mechanisms to make the use of resources compatible with their long-term conservation, and passed a series of laws that resulted in a National Sanitation Policy in 1967. In 1971, a pollution control program was started, and in 1972 there were around thirty-four federal organs that administered environmental policy, either directly or indirectly.[27] This shows that despite the most visible rhetoric adopted by the government (and also by most intellectuals of the period), Brazil was part of a global trend of environmental regulation.

Social mobilization played a crucial role in pressing the government for action, even though the environmental matters were marginal in the Brazilian public debate at the time.[28] From the mid-1970s on, grass-roots social movements, such as neighborhood associations, and several social movements of farmers, feminists and indigenous people, started to gain strength and visibility. In this sense the action of the Catholic Church, influenced by the Liberation Theology was very important because it acted as a protector of grass-roots organizations from the authoritarian state.[29] In this context, new environmentalist groups with explicit political agendas coming from an urban, middle-class context also emerged.[30] One example is the Community Heritage Defense Commission, formed by seventy groups from São Paulo state, which led a successful struggle against the construction of an airport in a forest next to the city of Caucaia. With the 1979 amnesty granted to political opponents in exile, arrived in Brazil the idea of a Marxism modified to embrace environmental issues, inspired by foreign green parties, especially around the figures of Fernando Gabeira and Alfredo Sirkis.

The environmental critique sprang also from the margins. Indeed, the urban poor, the peasant and indigenous population were the most excluded from the growth model as well as the industrial and development projects promoted by the regime but were also the ones that most suffered the environmental consequences: pollution, absence of sanitation, but also loss of land and livelihood. This situation led to important environmental grass-roots struggles such as one against the impact of pollution and poor living conditions led by slum populations of the heavily industrialized city Cubatão, named in the 1970s "the Valley of Death." The struggles of local communities also happened in far-flung regions of Brazil, typically over infrastructure or industrial projects encroaching their traditional spaces and land. This "environmentalism of the poor" was the beginning of what would become iconic grass-roots struggles in the 1990s such as the Rubber Tappers' Movement in the Amazon around the mythical figure of Chico Mendes or the Movement of People Affected by Dams (discussed further below).[31] The struggle of rubber tappers in the Brazilian state of Acre to prevent encroachment of cattle ranchers and land speculation in the mid-1970s started over an issue of livelihood. As the movement gathered support nationally and internationally it evolved into a fight for land rights combined with forest preservation creating the idea of extractive reserves. This quite revolutionary idea broke away from the traditional park conservation paradigm that excluded people, proposing instead conservation units in which the traditional population could continue to engage in their extractive and/or subsistence activity.[32]

In 1973, as a direct outcome of the Stockholm Conference, the Brazilian government created the *Secretaria do Meio Ambiente*, or Environment Secretariat (SEMA) to promote the conservation of the environment and rational use of natural resources. As many authors point out, this institutionalization was aimed at responding to international pressure[33] and was not equipped to carry out a real action of environmental protection, but rather to neutralize the opposition's criticism.[34] This shows, nonetheless, the influence of the Stockholm Conference and other international bodies and dynamics over this authoritarian government.[35]

On the Paraguayan side of the border, Stockholm seemed a place too far away to make an impact on domestic politics. Nevertheless, a locally brewed environmental consciousness also came to life in those years. A growing conservationist movement in the eastern borderlands on the subject of timber management helped to frame the issue in Paraguay. In the 1940s, a government journal, *La Cartilla Agropecuaria* (The Farmer's Journal) published an extensive article by agronomist Genaro Romero, an authoritative voice in issues related to the exploitation of national forest with vast experience in working for the government in matters related to the management of agricultural production. A product of earlier waves of conservationist movements in the United States, Romero was the representative of new thinking on forest management. For Romero, the depletion of natural forests had a double negative effect: on the one hand, it left the country short of crucial building resources and dependent on foreign provision with considerable financial costs; on the other hand, the regional climate worsened, endangering the "*bondad del clima local*" (the extremely mild local weather). Romero called farmers to avoid deforestation of useful trees and brought three crucial issues of the environmental agenda to local politics. First, he preached the "*explotación racional*" (rational exploitation) of national forests. Second, Romero called for massive reforestation policies across the country in order to avoid foreign dependency on natural resources. Third, he introduced the idea of acclimatization of exotic species to assure the success of reforestation policies.[36]

A more decisive driving force behind an inchoate environmental consciousness in Paraguay came from seeing the vanishing trees in the Eastern Region in the 1970s. As indicated above, Paraguay was a backwater of South America, with languid population growth in comparison with booming Brazil. By any standards, the human footprint in Paraguay lagged behind their Brazilian peers. However, that dormant development changed with the construction of the Itaipu Dam. The country grew at an average ratio of 8 percent of the annual GDP during the 1970s in the greatest economic jump of its history. Modernization came as the neighboring city of Puerto Presidente Stroessner grew overnight and even the capital city of Asunción was remodeled with money coming from Itaipu.[37]

The environmental politics of dam building in Brazil

Even though Itaipu was a partnership between ANDE and Eletrobrás, not only did the latter have greater experience in dam building but it also played the dominant role in the partnership. Paraguay had only built the Acaray dam before Itaipu, and it was a Brazilian firm (Mender Junior) that led the construction work. For that reason, the Brazilian electricity sector in the 1970s influenced decisively Itaipu's construction, and in that decade, with rising environment concerns around the globe, sustainability became a recurring issue in regional politics.

Reflecting on the environmental consciousness built in the previous decade, the Committee of the Electric Sector Company Management

(COGE) of the Eletrobrás published a working paper in 1989 assessing the introduction of environmental variables within the plans of development for the electricity sector.[38] The report brought up three external factors that proved decisive in the introduction of environmental concerns in the development plans of the sector. First, the sector incorporated society's pressure for the inclusion of environmental concerns in their development plans because of the collective political actions of environmental and social movements. Second, the sector designed development plans according to the emerging environmental laws passed by the government. Finally, the electricity sector was pressed by international funding organizations' environmental standards that forced new hydroelectric dams to conduct environmental impact studies. Internal factors also contributed to the growing environmental awareness of the electricity sector in Brazil. As many engineers reported, dam building transformed the environment so dramatically that the electricity sector could not turn a blind eye to the rising environmental issues that unfolded daily during their operations.

Brazilian national laws that regulated the impact of dam construction until the mid-1980s had mostly to do with the protection of the aquatic fauna in dam works, mainly through the establishment of hatcheries and fish farming in the reservoirs.[39] But during the 1970s, several policies and practices regarding the environment were developed in the electricity sector, responding to local issues such as water quality degradation, silting and erosion in the reservoirs, animal rescue, and others even if they did not prevent the adverse impacts of dam building. In this context, several companies were to create departments, work groups or divisions in charge of the "ecology" or the "environment."[40]

In this regard, Eletrobrás as coordinator of the electricity sector would play an important role. One of Eletrobrás's directors, the engineer César Cals, created in 1974 the Environmental Working Group (*Assessoria de Meio Ambiente*), to accompany the development of environmental studies and actions within the different companies that it supervised. The other objective of this group was to promote meetings around environmental issues, which would take place between 1976 and 1979.

As is often the case, the need for funding also played an important role in the adoption of environmental measures in dam building. Since international funding was critical for infrastructure projects during the military regime, the procedures, criteria, and demands of those agencies ended up being crucial in the planning and administration of public activities in Brazil and for its electricity sector in particular. Eletrobrás alone was, since 1960, the biggest borrower of the World Bank in Brazil, between 1949 and 1987 the World Bank loaned almost US$3.4 billion to the electricity sector in Brazil, reaching 10 percent of the bank's total investment in the Latin America and Caribbean region.[41]

As sources show, the World Bank (WB) imposed environmental assessment studies in the scope of dam projects being funded by the bank in

Brazil from the early 1970s.[42] This demand was made, for instance, in the frame of the loan for the Itumbiara dam (1973), the Sobradinho dam (1974), the São Simão dam (1972) and even the Salto Osório dam (1971), where issues over protecting the environment seem to appear for the first time.[43] However, the funding was not conditioned by those demands since they took place only once the project had already been approved. The World Bank's idea was rather to minimize the damage when it was deemed important.[44] Nonetheless, actors from the electricity sector perceived it as a requirement for the loans.[45] The Brazilian Ministry of Mines and Energy thus issued a directive to anticipate any potential demand.[46] In 1979, the National Department of Water and Electric Energy (DNAEE) established a directive, DG-009, that required environmental aspects to be addressed in the viability report in order to be approved. These reports were authored either by the American ecologist Robert Goodland[47] or by the São Paulo Electricity Company (CESP), which pioneered the development of environmental practices in Brazil.[48]

Regardless of the well-founded criticisms that those studies raised, the initiative was a first step toward the recognition of the environmental impact of dams and their deleterious effects on both human and non-human actors.[49] Nonetheless, the fact that Brazil remained under military rule in the 1970s limited the reach of these studies. When they did not receive the "restricted use" stamp from the government's intelligence service, their publication in English precluded Portuguese readers from reading it. Such was the case in the Tucuruí Dam. But if the regime was worried about the reactions to these impact studies, it was because the environment was gaining momentum within civil society and turning into a rationale for social mobilization. This was especially the case after 1977, when repression against activists decreased with the democratic opening started under the presidency of General Geisel.[50]

In fact, the first actors that confronted the electric sector, and therefore, the regime, were people directly affected by dam projects. In those days, infrastructure constructions were top-down projects directed by the military technocracy that was behind the modernization ideal. The population was forced to move from affected areas without much discussion, let alone explanation.[51] Often the only measure taken toward the affected population was a financial compensation of the owners of flooded land. The process of monetary compensation in Brazil was usually very complex, made more complicated in many regions since people did not even have land titles. Departments of relocation, which worked as real estate offices within the electricity companies, led the displacing of entire populations. This situation paved the way for unavoidable conflicts that increased as the affected populations mobilized locally and gathered supporters, and as their struggle became more intense and visible.[52]

Itaipu dam and environment

Even though several technical and diplomatic debates surrounded the construction of the Itaipu Dam, the process placed the environmental question and the social mobilization by grass-roots organizations at the center of the political agenda.[53] Even before the establishment of Itaipu Binacional, Robert Goodland had been hired to carry out an environmental assessment study for the construction of the Itaipu Dam in 1973. As the construction of the dam started, Itaipu Binacional launched in 1975 the "Basic Plan for Environment Conservation," which concerned the physical, biological, biotic, health, and cultural consequences of building Itaipu. In this framework were carried out archeological, climate, geological, epidemiological studies as well as an inventory of fauna and flora. This is considered the first holistic plan of environmental action for a Brazilian (or Paraguayan) dam. However, it was rather vague about the social impacts of the dam and how to deal with them, dedicating, for instance, only one paragraph to planning the zone's occupation by the displaced population. The plan, while well received in the press, could not keep at bay criticisms regarding deleterious effects for the environment that made the headlines—such as the risk of erosion and silting, the impact on local fauna, flora and native peoples, and the disappearance of the Guayrá/Sete Quedas Falls.[54]

On the Paraguayan side, the timber industry was benefiting from the boom in construction around the dam's works. House building for workers in the city near the dam region increased from 667 houses projected in 1975, to 3000 for the following two years. The growth of the timber industries accelerated deforestation in nearby forests in only two years of construction in the Itaipu Dam zone. By early June 1976, the Federación Paraguaya de Madereros (FPM, the timberloggers' union), created the same year, was already complaining about the brutal pace of deforestation without plans for reforestation to replace the used timber wood. Both the president of the FPM, Carlos Manuel Bedoya and the director of the Servicio Forestal Nacional (SFN) (National Forestry Secretary) called for joint private and public action to launch ambitious reforestation plans. Paraguay's 1973 Reforestation Law (Ley 422/73), which came in the year following the Stockholm Conference, did not help much to prevent more irresponsible damage to the forests of the Upper Paraná.

Soon after, and responding to these pressures from both countries, Itaipu launched the operation "Gralha Azul" (Portuguese for Blue Jay and named after a mythical bird of the Upper Paraná Region) in 1978 to reforest the lake's margins, and operation "Mymba Kuera" (Animals Rescue in Guarani) to save animals during the filling of the dam.[55] As noted by an engineer that worked on the Itaipu project, the concern was "how to do something that had to be impeccable for the international observers. There could be no faults, there could be no accidents, there could not be dead animals."[56]

The construction of the Itaipu Dam gave rise to an impressive protest movement.[57] The mainly peasant population, affected by the flooding of 1,350 km^2 between the cities of Iguaçu in Brazil and Ciudad del Este in Paraguay, mobilized on the Brazilian side in 1978 with the help of the local church linked to the Comissão Pastoral da Terra (CPT). The population worried about low compensation payments, the lack of information about the construction and the collapse of their communities. Itaipu Binacional refused to negotiate, and the people affected organized meetings, petitions, and street demonstrations, gathering support from other peasant movements, local opposition parties, and environmentalists.[58] In April 1979, an assembly of 3000 people was held, and the movement was baptized 'Justiça e Terra' (Justice and Land). After that, the company decided to negotiate. Famous poets such as Carlos Drummond de Andrade, also stood with the resistance movement when he protested the flooding of the Guayrá/Sete Quetas Waterfalls.[59]

On July 14, 1980, a political demonstration took place in front of the Itaipu's headquarters in Santa Helena, in the Brazilian State of Paraná, with significant repercussions in the press. The population affected by Itaipu camped there for seventeen days, and during that time they received support from other movements, as well as clerics from the Catholic Church and members of the parliament. The company agreed to pay 80 percent of the value asked by the movement. In 1981, the population camped again for fifty-four days on the road that led both to Itaipu's headquarters in Foz do Iguaçu and to Paraguay and renamed it the "*trevo da vergonha*" (intersection of shame). After this action, the company agreed to pay the full indemnities asked by the movement and gave two areas for people's resettlement.[60]

Likewise, the indigenous Avá Guarani population also mobilized with the support of the Catholic Church, the World Bank and the Brazilian Association of Anthropology and managed to secure an area of 253 hectares. Ultimately, they were settled on a farm of 1.7 thousand hectares purchased by the government in 1997.

The legacy of Itaipu: the impact of the dam on environmental regulation across the borders

As pointed out by Germani, "the new fact presented by the expropriated of Itaipu was not the resistance in itself, but the organized resistance."[61] The way the company treated the population in Itaipu served as an example in organizing the populations that were affected by other dam projects in Brazil. The most successful example was the mobilization of the Regional Commission of Affected People by the Dams in High Uruguay (CRAB), where in 1979 the construction of twenty-five hydroelectric dams was planned. Leaders of CRAB visited the Itaipu Dam and its resettlement project and met several times with the leaders of social groups affected by Itaipu.[62] As in the case of the resistance movement against Itaipu, CRAB relied on ecclesiastical organizations and the peasant movement to organize the populations, situated throughout the zone, who were threatened by projects where they created several local committees.

The strategy proved to be effective. In February 1983, the movement organized a demonstration, the "*Romaria da Terra*" (Pilgrimage of the Earth) with the slogan "Waters for life and not for death," in which 20,000 people participated. In 1985, the movement handed to the government the petition "No to the Dams" which had more than a million signatures. The social organization succeeded in hampering the development of the projects (such as the Machadinho dam and Itá) and compelled the company (here the Eletrosul) to negotiate with the local commission.

The resistance of the population affected by Itaipu also had influence beyond Brazil. This was the case during the construction of Yacyretá Dam further south along the Paraná River in the borderlands between Argentina and Paraguay, where the population also organized a struggle against dam builders. As explored by anthropologists Carmen Ferradas and Gustavo Lins Ribeiro, people in both the Paraguayan city of Encarnación and the Argentinean city of Posadas organized and successfully set better terms of compensation for the damage caused by the construction of the Yacyretá Dam.[63]

Despite the government's violence, repression, and the asymmetrical relations between those affected by the dams and the electric companies, social movements built a growing environmental coalition that involved leftist political parties, civil society, and public opinion, forcing the government and the electricity sector to negotiate, and in some cases, even block dam construction altogether. These victories came in the form of better compensation or better relocation plans, but also in changes in the electricity sector plans that better reflected the interests of local actors. As indicated by energy sector specialists in Brazil, if there were any advancements in the social domain, it was "*na marra*" (forcefully), even if the results of these actions were ambiguous: "I don't want to say that the rights are recognized, but they [the affected population] are recognized as having rights."[64] As Eletrobrás itself admits, the actions of groups such as CRAB "on the government's organs, and on the public opinion, made companies carry out, progressively, actions that are more responsible, regarding environmental problems that result from their activities."[65]

Environmental resistance against dam builders gained momentum after the demise of the military dictatorship in 1984 in Brazil and in 1989 in Paraguay. During the first National Meeting of Workers affected by dams in April 1989, the participants created the Movement of People Affected by Dams. This organization still leads the efforts to fight against old and new dam projects to this day.[66]

The environmental institutionalization and action of the energy sector in the 1970s influenced legislation on the environment passed in Brazil[67] and in Paraguay[68] and the Itaipu's experience was crucial in this process.

One example is the concept of a biological refuge zone destined to receive wild animals displaced by infrastructure projects, which was introduced in Brazil during the Itaipu experience following the Mymba Kuera rescue operation.[69] Also Itaipu's environmental control plan is considered the first one of this kind to be made in Brazil or in Paraguay and subsequently influenced the practice of the whole infrastructure construction sector. When the National Environmental Policy (PNMA) was approved in Brazil in 1983, the obligation to have environmental permits, environmental impact studies and government offices to carry them out in the framework of development projects was one of the main victories. That practice had been outlined by the energy sector, which had informally carried out impact studies since the early 1970s and where the Itaipu's study was one of the first. This practice became law in 1979 with the DNAEE's environment assessment studies directive.

In the Paraguayan case, the Itaipu experience modeled the future negotiation of the construction of the Yacyretá Dam down the Paraná River with Argentina. The entire law design, put in practice with Itaipu, became the standard for the future construction of Yacyretá. Furthermore, many of the problems unforeseen during the construction of the Itaipu Dam, were considered before embarking on the construction of the Yacyretá Dam. The role of the Itaipu Dam as nature's steward reshaped political institutions in Paraguay. Most of the government offices such as the Ministry of Agriculture and Cattle Raising (MAG), the Office for the Protection of Forests (INFONA), and the Vice Ministry of Energy and Mining (VMME) not only followed Itaipu's guidelines to manage natural resources but also recruited former personnel of Itaipu to occupy high level positions in the 1970s, 1980s, and 1990s.

Conclusion

This chapter has tried to show how two Latin American dictatorships faced the new framing of the environment in the 1970s, and how dam building participated in shaping the responses. Brazil and Paraguay, two very different countries, had a long history of war and border conflicts. But as both countries were under the rule of military governments in the second half of the twentieth century they embarked together in the construction of the biggest infrastructure project in South America at the time, the Itaipu Dam. The two countries practiced the national-developmentalist strategy and put forward economic growth based on the idea of exploiting natural resources to fuel development.

Before the 1970s, dams had been built without much concern for the extreme transformation it entailed on the nature, landscape, and lives of the humans and non-humans they affected. The impacts of these infrastructure projects were seen as largely positive (economic growth, urban electrification, industrial hubs) and the negative impacts were masked, seen as minor issues or even as necessary sacrifices.[70] Nonetheless, the environmental consciousness that emerged in the 1970s caught up with the Paraguayan and Brazilian regimes and also with their electricity sector, demanding corrective action by the governments.

This chapter has shown how through international pressure, notably in the framework of the United Nations' Stockholm Conference in 1972, both countries created legal frameworks and institutions to deal with the environment. But the action of environmental and social movements, especially grass-roots organizations as the Itaipu's case reveals, proved to be much more effective in pushing the government to take action on the environment. The trajectory of these mobilizations during the military regime cross-pollinated with other popular social movements of the time, such as the peasant organizations and trade union movements that relied on the same resistance toolkit with manifestos, petitions, meetings, assemblies, demonstrations, and a strong mediating presence of the Catholic Church. Regardless of the lasting image of those dictatorships as anti-environmentalist, as our article shows, by the end of those military regimes, Brazil and Paraguay had an important framework of environmental laws and institutions. Unfortunately, those frameworks were not always put into practice.[71]

Notes

1 Ligia Maria Martins Cabral and Eliane Rocha Correia de Azevedo, *O meio ambiente e o setor de energia elétrica brasileiro* (Rio de Janeiro: Centro da Memória da Eletricidade no Brasil, 2009), 59–61; Richard Niland, "Death by Water, The Rise and Fall of Los Saltos del Guairá," *Environmental History* 23 (2018): 56–81.

2 ECLAC's researchers Raúl Prebisch and Celso Furtado observed that the classical theory of growth could not be applied to underdeveloped countries since it was born within the framework of the observation of social structures of industrialized societies, what Prebisch called the "false universalism" of economic science. Underdevelopment for them had to be understood in the unequal relationship between countries in "the center" and in "the periphery," so they proposed that growth should be directed to the internal market rather than external market through the development of an import substitution industry of consumption goods. José Maria Dias Pereira, "Uma breve história do desenvolvimentismo no Brasil," *Cadernos do Desenvolvimento* 6, no. 9 (December 2011): 121–141.

3 Luis Carlos Bresser-Pereira, "O Estado na Economia Brasileira," *Ensaio de Opinião* 4, no. 2–2 (1977): 16–23; Luis A. Campos, *Apuntes de historia economica del Paraguay: desarrollo, auge y decadencia de una economia de enclaves* (Asunción: Intercontinental Editora, 2016).

4 Antonio Carlos Bôa Nova, *Energia e classes sociais no Brasil* (São Paulo: Edições Loyola, 1985): 142; Alfredo Stroessner, *Acaray: Centennial of the National Ordeal* (Asunción: S.I., 1968).

5 Centro da Memória da Eletricidade no Brasil, *Panorama Do Setor de Energia Elétrica No Brasil* (Rio de Janeiro: Centro da Memória da Eletricidade no Brasil, 1988); Myrna Santiago, *The Ecology of Oil: Environment, Labor, and the Mexican Revolution, 1900–1938* (Cambridge and New York: Cambridge University Press, 2009); Fernando Coronil, *The Magical State: Nature, Money, and Modernity in Venezuela* (Chicago, IL: University of Chicago Press, 2008).

6 Data from the Historical Electricity Installed Capacity Statistics Dataset from "The Shif Project." www.tsp-data-portal.org. Accessed: May 2, 2018.

7 Bôa Nova, *Energia e classes sociais no Brasil*, 113.

8 José Luiz Lima. *Estado e energia no Brasil: o setor elétrico no Brasil: das origens à criação da Eletrobrás (1890–1962)* (São Paulo: IPE/USP, 1984), 115; Oliver J. Dinius, *Brazil's Steel City: Developmentalism, Strategic Power, and Industrial Relations in Volta Redonda, 1941–1964* (Stanford, CA: Stanford University Press, 2011).

9 *Bandeirantes* were Portuguese explorers who, since the seventeenth century, raided Brazilian territory in search of Indian slaves and riches. Richard M. Morse (ed.), *The Bandeirantes: The Historical Role of the Brazilian Pathfinders* (New York: Alfred A. Knopf, 1965).

10 Thomas L. Whigham and Barbara Potthast, "The Paraguayan Rosetta Stone: New Insights into the Demographics of the Paraguayan War, 1864–1870," *Latin American Research Review* 34, no. 1 (1999): 174–186; and Hendrik Kaay and Thomas Whigham, *I Die with my Country, Perspectives on the Paraguayan War* (Lincoln, NE and London: University of Nebraska Press, 2004); and Leslie Bethell, *The Paraguayan War (1864–1870)* (London: The University of London, Institute of Latin American Studies, Working Papers, 1996). For recent scholarship on the War see Michael Kenneth Huner, "*Toikove Ñane Retã!* Republican Nationalism at the Battlefield Crossings of Print and Speech in Wartime Paraguay, 1867–1868," in W. G. Acree Jr. and J. C. González Espitia (eds.), *Building Nineteenth-Century Latin America: Re-rooted Cultures, Identities, and Nations* (Nashville, TN: Vanderbilt University Press, 2009), 79–97.

11 For general descriptions of the relations between the two countries see Thomas Whigham, *The Paraguayan War* (Lincoln, NE: University of Nebraska Press, 2002); Andrew R. Nickson, "Brazilian Colonization of the Eastern Border Region of Paraguay." *Journal of Latin American Studies* 13, no. 1 (1981): 111–131; Alfredo Da Mota Menezes, *La herencia de Stroessner* (Asunción: Carlos Shauman, 1990); Domingo Laíno, *Paraguay: fronteras y penetración brasileña* (Asunción: Cerro Corá, 1977); Jacob Blanc, "Itaipu's Forgotten History: The 1965 Brazil–Paraguay Border Crisis and the New Geopolitics of the Southern Cone," *Journal of Latin American Studies* (2017): 1–27; Melissa Birch, "La política pendular: Política de desarrollo del Paraguay en la post-guerra," in *Política exterior y relaciones internacionales del Paraguay contemporáneo*, José Luis Simón (comp.) (Asunción, Centro Paraguayo de Estudios Sociológicos, 1990); Francisco Doratioto, *Una relación compleja: Paraguay y Brasil 1889–1954* (Asunción: Tiempo de Historia, 2011); Paul H. Lewis, *Paraguay Under Stroessner* (Chapel Hill, NC: University of North Carolina Press, 1989).

12 DG no. 591, July 7, 1926; DG no. 603, July 9, 1926; and DG no. 1042, December 10, 1926; Sección Política y Diplomática, no. 190, año 1926, Archivo del Ministerio de Relaciones Exteriores, República del Paraguay.

13 Letter from Merwin L. Bohan, President US Section, Joint Brazil—United States Economic Development Commission, September 3, 1952, Box 33, Entry UD 3088, National Research Archives.

14 Osny Duarte Pereira, *Itaipu, pros e contras* (Rio de Janeiro: Editora Paz e Terra, 1974), chapter 4; Melissa H. Birch, "Public Enterprise and Economic Development: The Case of ANDE in Paraguay" (PhD Thesis, The University of Illinois at Urbana-Champaign, IL), 68–74 and 90; John Howard White, chapter II; Enzo Debernardi, *Apuntes para la Historia Política de Itaipú* (Asunción: Editorial Gráfica Contínua, 2010), 78–84; Comissão Interestadual da Bacia Paraná-Uruguai, *Análise Preliminar do Desenvolvimiento Economico da Bacia Paraná-Uruguai* (São Paulo: Serviços de Planejamento, 1956), 41–43.

15 Dispatch No. 245, November 19, 1956, Memorandum of Conversation, p. 2 Box 31, Entry UD 3088, National Research Archives.

16 "Stroessner faz acôrdo com Goulart: Sete Quedas," *Jornal do Brasil*, January 21, 1964, 1.

17 No. 319/954, October 6, 1954; No. 338/953, August 8, 1953, Departamento de Política Diplomática, no. 486, Archivo del Ministerio de Relaciones Exteriores, República del Paraguay.

18 One example is the idea of developing nuclear energy in Paraguay, following the model of the Brazilian plan, which had been under development since the Kubitschek government. Delegations from Paraguay went to Brazil, such as in a 1964 visit to the Brazilian National Commission of Nuclear Energy, and the other way around. If Paraguay did not end up developing a nuclear program, it followed Brazil's steps at least by reproducing similar bureaucratic offices.

19 Alfredo Da Mota Menezes, *A Herança de Stroessner, Brasil-Paraguay 1955–1980* (Campinas: Papirus, 1987), 57–58; R. Andrew Nickson, "The Itaipú Hydro-Electric Project: The Paraguayan Perspective," *Bulletin of Latin American Research* 2, no. 1 (October 1982), 3–5; Efraím Enríquez Gamón, *Itaipú, Aguas que Valen Oro* (Asunción: ABC Color, 2007), 41. On Ciudad del Este, see Edgar L. Ynsfrán, *Un Giro Geopolítico: el milagro de una ciudad* (Asunción: Instituto Paraguayo de Estudios Geopolíticos e Internacionales, 1990); Alfredo M. Seiferheld. *Nazismo y fascismo en el Paraguay. Volume 2, Los años de la guerra, 1939–1945* (Asuncion: Histórica, 1986), 132; and Fidel Miranda Silva, *Historia del Alto Paraná* (Ciudad del Este: AGR, 2007), 24.

20 Robert T. Daland, *Brazilian Planning: Development Politics and Administration* (Chapel Hill, NC: University of North Carolina Press, 1967); Wilson Quintella, *Memorias do Brasil Grande: a história das maiores obras do país e dos homens que as fizeram* (São Paulo: Editora Saraiva, 2008); and Anibal Miranda, *Paraguay y las obras hidroeléctricas binacionales* (Asunción: El Lector, 1988).

21 The shift of Paraguayan politics from the long-lasting Argentinean predominance to the Brazilian orbit constituted a significant historical rupture. In 1943, Paraguayan president Higinio Morínigo claimed, after a trip to Argentina, that "history and geography brought together the two countries through blood and cordiality into a shared destiny of firm and indestructible bonds today stronger than ever." *El Paraguayo*, Asunción, December 19, 1943; J. Eliseo Da Rosa, "Economics, Politics, and Hydroelectric Power: The Paraná River Basin," *Latin American Research Review* 18, no. 3 (1983), 80–81; Diego Abente, "Constraints and Opportunities: Prospects for Democratization in Paraguay," *Journal of Interamerican Studies and World Affairs* 30, no. 1 (Spring 1988), 74–77; Werner Baer and Melissa Birch, "The International Economic Relations of a Small Country: The Case of Paraguay," *Economic Development and Cultural Change* 35, no. 3 (April 1987), 67.

22 José Costa Cavalcanti, interviewed by *Journal of the Ilha Grande*, cited by Luis Eduardo Catta, *O Cotidiano de uma Fronteira, A perversidade da Modernidade* (Cascavel: Edunioste, 2000), 71. The argument proves compelling even now, as two of the most virulent critics of the project, Paulo R. Schilling and Ricardo Canese are still only arguing about the unfairness of the Itaipu Treaty and the distribution of the earnings rather than presenting an alternative view on the management of nature. Paulo R. Schilling and Ricardo Canese, *Itaipu, Geopolítica e Corrupçao* (São Paulo: CEDI, 1991), 20 and 40–41. The idea of God could also be found in critical views of the project such as German pastor Gernote Kirinus who, after unsuccessfully dealing with the Itaipu's technocrats for the concession of reparations to peasants displaced by the floods in 1978, stated that "[dealing with Itaipu's officials] was like negotiating with a new miraculous God, a technocrat who thinks that any issue can be solved through the magical touch of

technique," cited in Juvencio Mazzarollo, *A Taipa da Injustiça, Esbanjamento economico, drama social e holocausto ecologico em Itaipu* (Curitiba: Edições Loyola, 2003), 61.

23 Antonio Teixeira de Barros, "Brazil's Discourse on the Environment in the International Arena, 1972–1992," *Contexto Internacional* 39, no. 2 (May/August, 2017): 421–442. For a general account of the conference see Avi Brisman, Stockholm Conference, 1972, in D. K. Chatterjee (ed.), *Encyclopedia of Global Justice* (Dordrecht: Springer, 2011).

24 Eduardo Viola and H. R. Leis. "A evolução das políticas ambientais no Brasil, 1971–1991: do bissetorialismo preservacionista para o multissetorialismo orientado para o desenvolvimento sustentável," in *Dilemas socioambientais e desenvolvimento sustentável* (Campinas: Editora da Universidade Estadual de Campinas, 1992).

25 Philippe Le Prestre, *The World Bank and the Environmental Challenge* (Selinsgrove: Susquehanna University Press, 1989).

26 Some authors, like Sánchez, argue that it was the first time Brazil had an environmental policy per se, since the older laws pertaining to this sphere like the Water, Forest and Mining Code (all passed in 1934), were simply legal instruments that aimed at rationalizing the use and exploitation of natural resources through sectoral policies in the framework of the consolidation of Brazilian capitalism in a context of early industrialization. Inversely, other authors such as Augusto Pádua point to a much earlier emergence of an ecologist view in Brazil that echoed the emergence of ideas of ecosystems in Europe. Pádua's research shows that at the end of the eighteenth century in Brazil, a "proto-environmentalism" emerged in intellectual circles debating deforestation and land degradation. In 1934, the first Conference of Nature Protection was held in Rio de Janeiro and many laws that regulate natural resources use and protection are adopted (the above-mentioned, water, forest, and mining code) and National Parks came into being, like the Itatiaia Park in Rio de Janeiro in 1937. Under this light, the environmental legislation that emerged during the military regime is but one phase of a longer process of State policy and action on the environment. José Augusto Pádua, *Um sopro de destruição: pensamento político e crítica ambiental no Brasil escravista, 1786–1888* (Rio de Janeiro: Jorge Zahar Editor, 2002); José Augusto Pádua, "Environmentalism in Brazil: A Historical Perspective," in J. R. McNeill and E. Maldin (eds.), *A Companion to Global Environmental History* (Oxford: Wiley-Blackwell, 2012), 455–473; Luis Enrique Sánchez, *Avaliação de impacto ambiental: conceitos e métodos* (São Paulo: Oficina de Textos, 2008). For a broader take on the Upper Paraná Region including national parks in the frontier region see Fred Freitas and Jacob Blanc, *Big Water: The Making of the Borderlands Between Brazil, Argentina and Paraguay* (Tucson, AZ: Arizona University Press, 2018); and Fred Freitas, "A Park for the Borderlands: The Creation of the Iguaçu National Park in Southern Brazil, 1880–1940," in *Revista de Historia Iberoamericana*, 2014. For the case of Mexico see Emily Wakild, *Revolutionary Parks: Conservation, Social Justice, and Mexico's National Parks 1910–1940* (Tucson, AZ: University of Arizona Press, 2011).

27 Kathryn Hochstetler and Margaret E. Keck. *Greening Brazil: Environmental Activism in State and Society* (Durham, NC: Duke University Press, 2007).

28 Eduardo Viola, 1987, op. cit.

29 Liberation Theology, a theological branch of the Catholic Church born in Latin America after the Second Vatican Council, considered a Christian duty to fight for the poor and oppressed in the search for a better world.

30 Hochstetler and Keck, op. cit. and Pádua, 2012, op. cit.

31 Juan Martinez-Alier, *The Environmentalism of the Poor* (New York: Oxford University Press, 2005).

32 For a broader take on the Brazilian rubber boom see Barbara Weinstein, *The Amazon Rubber Boom 1850–1920* (Stanford, CA: Stanford University Press, 1985). For the story of Chico Mendes, see Mary Nakashima, *Chico Mendes por ele mesmo* (São Paulo, Brazil: Martin Claret, 1992).

33 Eduardo Viola, "O Movimento ecológico no Brasil (1974–1986): Do ambientalismo à ecopolítica," in J. A. Pádua (ed.), *Ecologia e política no Brasil* (Rio de Janeiro: Editora Espaço e Tempo, 1987); Luiz C. Barbosa, "The 'Greening' of the Ecopolitics of the World-System: Amazonia and Changes in the Ecopolitics of Brazil," *Journal of Political & Military Sociology* 21, no. 1 (Summer 1993): 107–134.

34 Eduardo Viola, "O Movimento ecológico no Brasil (1974–1986): Do ambientalismo à ecopolítica," in J. A. Pádua (ed.), *Ecologia e política no Brasil* (Rio de Janeiro: Editora Espaço e Tempo, 1987); Luiz C. Barbosa, "The 'Greening' of the Ecopolitics of the World-System: Amazonia and Changes in the Ecopolitics of Brazil," *Journal of Political & Military Sociology* 21, no. 1 (Summer 1993): 107–134; P. Roberto Guimarães, "Ecologia e Política Na Formação Social Brasileira," *Revista de Ciências Sociais* 31, no. 2 (1988): 263.

35 David John Frank, Ann Hironaka, and Evan Schofer, "The Nation-State and the Natural Environment over the Twentieth Century," *American Sociological Review* 65, no. 1 (2000): 96–116, show a global tendency of creation of those types of structure since the institutionalization of the environment in the international sphere has a strong influence on the countries that participate in those stances.

36 *El País*, Asunción, Paraguay, November 23, 1942, p. 2.

37 R. Andrew Nickson, "The Itaipu Hydro-Electric Project: The Paraguayan Perspective," *Bulletin of Latin American Research* 2, no. 1 (October, 1982): 1–20.

38 Eletrobrás (COGE), *Os serviços de energia elétrica e o meio ambiente COGE SJU 17–89* (Rio de Janeiro: Eletrobrás, 1989), 4.

39 The very first legal obligation in Brazil that deals with the milieu impacted by dams was São Paulo State law 2250 from December 28, 1927. Later the Water Code from 1934 and the Fishing Code of 1938 made the obligation to protect fish in the construction of dams a national policy. Those fish protection programs continue to be promoted during the military government with the decree-law n°221 of 1967 and through a 1971 directive from Superintendence for the Development of Fishery (SUDEPE) that implement a national program for the protection of aquatic wildlife in the reservoirs of Brazilian dams.

40 In 1983, a survey from Eletrobrás identified that fourteen out of thirty-three electricity companies developed some kind of environment protection action. Cabral, Azevedo, op. cit., 142.

41 Stephan Schwartzman and Michelle Malone, "Os bancos multilaterais de desenvolvimento e o setor energético," in Leinad Ayer O. Santos and Lúcia M. M. de Andrade (eds.), *As Hidrelétricas do Xingu e os povos indígenas* (São Paulo: Comissão Pró-Indio de São Paulo, 1988), 68.

42 Even if several authors point out that it is only in 1987 that the Bank starts to integrate the environment component in the project cycle (through the requirement of environmental impact studies) it had, since 1972, an Office of Environmental Adviser in charge of analyzing the potential impacts of financed projects. Le Prestre, op. cit., 125.

43 World Bank, *Brazil—Salto Osório Hydroelectric Plant Project*, Public Utilities Projects no. PU 59 (Washington, DC: World Bank, 1971); World Bank, *Brazil—São Simão Hydroelectric Project* (Washington, DC: World Bank, 1972); World Bank, *Brazil—Fourth Paulo Afonso Hydroelectric Power Project* (Washington, DC: World Bank, 1974); World Bank, *Brazil—Itumbiara Hydroelectric Power Project* (Washington, DC: World Bank, 1973).

44 Le Prestre, op. cit., 26–27.

45 Fernando Bidegain, *[depoimento 2003]* (Rio de Janeiro: Centro de Memória da Eletricidade no Brasil, 2003).
46 Renato Feliciano Dias (ed.), *Ciclo de palestras: a Eletrobrás e a história do setor de energia elétrica no Brasil* (Rio de Janeiro: Centro da Memória da Eletricidade no Brasil, 1995).
47 From the Cary Arboretum of the Botanical Garden of New York but who had important ties with Brazil where he had done fieldwork for his PhD in tropical ecology.
48 Goodland was alone responsible for seven out of seventeen "environmental assessment"-like studies carried out before legislation made it an obligation. Seven others were done by the CESP many in partnership with famous architect Ruy Ohtake.
49 For some examples of critical analysis of the military regime period impact assessment studies see Elizabeth Monosowski, *Tucuruí Dam in the Amazon: Development at Environmental Cost?* (Berlin: IIES, 1984); Philip M. Fearnside, "Brazil's Balbina Dam: Environment versus the Legacy of the Pharaohs in Amazonia," *Environmental Management* 13, no. 4 (July 1, 1989): 401–423; Robert E. Stein and Brian Johnson, *Banking on the Biosphere?* (Lanham, MD: Lexington Books, 1979); Luiz Pinguelli Rosa et al., *Impactos de grandes projetos hidrelétricos e nucleares: aspectos econômicos e tecnológicos, sociais e ambientais* (São Paulo, SP: Marco Zero, 1988).
50 Hochstetler and Keck, op. cit., 32.
51 This process is narrated by several case studies of dam constructions during the military regime and before. See, for instance, Carlos B. Vainer, "Águas Para a Vida, Não Para a Morte: Notas Para Uma História Do Movimento de Atingidos Por Barragens No Brasil," in H. Acselrad, S. Herculano, and J. A. Pádua (eds.), *Justiça Ambiental e Cidadania* (Rio de Janeiro: Relume Dumará, 2003) and Sigaud, Lygia, *Efeitos sociais de grandes projetos hidrelétricos as barragens de Sobradinho e Machadinho* (Rio de Janeiro: Museu Nacional/PPGAS, 1986).
52 As some authors underline, the emergence of a dissent movement brought up an unavoidable but unforeseen "impact" of dam building during the decline of the dictatorship. Carlos B. Vainer, "Águas Para a Vida, Não Para a Morte: Notas Para Uma História Do Movimento de Atingidos Por Barragens No Brasil," in H. Acselrad, S. Herculano, and J. A. Pádua (eds.), *Justiça Ambiental e Cidadania* (Rio de Janeiro: Relume Dumará, 2003); M. M. Cernea. "Involuntary Resettlement: Social Research, Policy and Planning," in M. M. Cernea (ed.), *Putting People First: Sociological Variables in Rural Development* (New York: Oxford University Press for the World Bank, 1991), 188–215.
53 Juvencio Mazzarollo, *A Taipa da Injustiça, Esbanjamento economico, drama social e holocausto ecologico em Itaipu* (Curitiba: Edições Loyola, 2003).
54 "Itaipu divulga plano para preservar o meio-ambiente," *O Estado de S. Paulo*, October 16, 1974, p. 36; "Usina de Itaipu vai afetar o meio-ambiente," *O Estado de S. Paulo*, July 21, 1974, p. 22; "Itaipú: los Ava Chiripá aún no tienen a donde ir," *Sendero*, Week 1, Number 223, April 1982; "Itaipú: los Avá Chiripá, Porqué oponer el progreso a la justicia?," *Sendero*, Number 222, March 19, 1982.
55 Cabral and Azevedo, op. cit., 62.
56 Eurico Miller, *[depoimento 2003]* (Rio de Janeiro, Centro da Memória da Eletricidade no Brasil, 2003). For a larger take on Brazil's fixation with progressive regulation, impressive for foreigners but with only limited or even null local effect see John D. French, *Drowning in Laws: Labor Law and Brazilian Political Culture* (Chapel Hill, NC: University of North Carolina Press, 2004).
57 On this contestation movement see Guiomar Inez Germani, *Expropriados: terra e água: o conflito de Itaipu* (Salvador: EDUFBA/ULBRA, 2003).
58 Germani, op. cit.

59 He will write the poem "Adeus as Sete Quedas" and several other articles against the destruction of the Seven Falls in Jornal do Brasil in 1982.
60 Cabral and Azevedo, op. cit., 65.
61 Germani, op. cit., 99.
62 Vainer, op. cit.
63 Carmen Ferradas, *Power in the Southern Cone Borderlands: An Anthropology of Development Practice* (Wesport, CT: Berbin and Garvey, 1998); and Gustavo Lins Ribeiro, *Transnational Capitalism and Hydropolitics in Argentina: The Yacyretá High Dam* (Gainsville, FL: University of Florida Press, 1994).
64 Paula Yone Stroh, *[depoimento 2003]* (Rio de Janeiro: Centro da Memória da Eletricidade no Brasil, 2003).
65 Eletrobrás (COGE), op. cit., 4.
66 For more on the subject of the creation of the MAB see: Ilse Scherer-Warren, "Do local ao global: a trajetória do Movimento dos Atingidos por Barragens (MAB) e sua articulação em redes," in F. Rothman, R. Ribeiro, and A. Zhouri (eds.), *Vidas Alagadas: Conflitos Socioambientais, Licenciamento e Barragens* (Viçosa: Editora UFV, 2007); Vainer, op. cit.
67 Namely the National System for the Environment (SISNAMA) in 1981, the National Environmental Policy (PNMA) in 1983, and the National Environmental Council (CONAMA) in 1984.
68 Such as the National Reforestation Law of 1973 (Ley 422/73).
69 Cabral and Azevedo, op. cit., 211.
70 Centro da Memória da Eletricidade no Brasil, op. cit.
71 "Legislação Federal Sobre o Meio Ambiente: Uma Avaliação Crítica," in E. Moreira Oliveira and Z. Kacowicz (eds.), *Relatório da Qualidade do Meio Ambiente (Sinopse)* (Brasília: Ministério do Interior and SEMA, 1984).

5 Crafting authoritarian atmospheres under Pinochet's dictatorship

Leonardo Valenzuela Pérez

Introduction

The Chilean military dictatorship begun by the September 1973 coup and led by General Augusto Pinochet is infamous for its sinister record of political violence and human rights violations. Alongside this bloody legacy, the regime is also well known as an early implementer of structural reforms guided by an orthodox reading of neoliberalism, and authoritarian liberalism more broadly, based on principles of strong state and free economy.[1] Its agenda entailed paring down workers' rights, privatizing public utilities and corporations, and dismantling welfare institutions, all of which served to exacerbate already stark socioeconomic inequalities. The regime's model of economic development, which relied heavily on foreign investment for the extraction and export of natural resources, accumulated a massive environmental toll during its seventeen years in power, characterizing a particular brand of right-wing authoritarian environmentalism heavily focused on the commodification of nature during the last decades of the Cold War.

It is a commonplace in Chile to assume that environmental policy was an innovation brought about by democratic governments after the end of Pinochet's dictatorship in 1990. The emergence of environmentalism as a new and progressive social movement, aligned with Pinochet's opposition, in the 1980s contributed to that perception. Some popular narratives even suggest that the General Environmental Framework Law (N°19.300) of 1994 was a direct reaction to the authoritarian regime's incognizance and inattention to environmental matters. However, recent literature has made clear that, despite the dictatorship's poor record on the environment, post-1990s democratic governments' approaches to environmental policy proceeded from the policy debates and initiatives adopted during the dictatorship.[2] Far from representing a rupture, environmental policy in the 1990s was profoundly shaped by the ideas and actions of regime-era agents, arguably constituting a continuity in doctrine and methods.

The dictatorship's approach to environmental protection was epitomized by the 1980 constitutional right to live in a pollution-free environment, the enforcement of which was nearly impossible from a technical perspective and lacked clear legal mechanisms and political will. Regime officials typically saw environmental protection and regulation as a threat to economic development. In practice, they were more concerned with inventing and testing new market-based mechanisms to handle environmental issues without state intervention, according to neoliberal precepts, rather than solving environmental problems with existing methods that were known to be effective through the direct state regulation of industrial and other economic activities. Finally, despite its lack of accountability to democratic principles, the regime's environmental policy decisions were particularly susceptible to international diplomatic pressures and local industrial interests, as has been the case with dictatorships elsewhere.

A central environmental concern for the dictatorship was air pollution, particularly pollution in the expanding metropolitan area of Santiago and pollution derived from the operation of copper smelters, a central industry for Chilean economic development. Government awareness of air pollution pre-dated Pinochet's regime and had been part of successive governments' agendas since the early 1960s.[3] Despite that early awareness and policy responses, more than five decades later, urban and mining air pollution are still the top environmental issues faced by the Chilean state.[4] This chapter considers the iconic case of Quintero Bay on Chile's central coast, the longest-standing environmental conflict in Chilean history, to illustrate the complex operations of environmental policy and politics during the Pinochet era. The toxic legacy of the dictatorship is still visible today: Quintero Bay is today one of the most prominent cases of chronic pollution in Chile after more than fifty years of environmental catastrophe and conflicts that are still far from resolution. The argument presented here is that the dictatorship's technical and ideological approaches not only worsened toxicity in Quintero Bay but also created precedents that would continue to characterize the approach of the Chilean government to the environment long after the transition to democracy.

The chapter argues that at least three sets of factors intervened in the overall crafting of air pollution conditions of Quintero Bay: first, the position of Chile in the Cold War as well as sanctions following the military coup created technological and trade dependencies unconducive to environmental protection; second, the ideological orthodoxy of neoliberal technocrats in charge of public sector investments favored experimental market mechanisms that were ineffective for environmental regulation; and third, the supposed de-politicization of environmental issues, through the introduction of the right to a pollution-free environment in the new Political Constitution of 1980, produced intractable conflicts with industrial and economic development interests.

This chapter seeks to contribute to the study of historical practices of environmental protection and destruction in Cold War era authoritarian regimes, excavating the dynamics of Pinochet's dictatorship with an emphasis on its environmental practices and outcomes. The analysis is centered on the reshaping of air pollution as a policy concern for the dictatorship, using the case of Quintero Bay. The materials used for this chapter come from extensive archival work at the Chilean National Archive of the Administration, the Chilean National Service of Geology and Mining (SERNAGEOMIN) library, and the Proceeding of the Government Junta Sessions published by the Chilean National Library of Congress.

Approaches to studying dictatorships and the environment

The crisis and fall of the Soviet Union was accompanied by publications that emphasized the environmental toll of Eastern Bloc regimes, portraying environmental destruction as yet another consequence of Socialism and in many ways overlooking the environmental consequences of militarism and industrial development in the West.[5] Since then, the study of the relationship between authoritarianism and the environment has been marked by theoretical proposals linking dictatorships with the capacity to impose social values aligned with environmental protection and take decisive action against threats like climate change. The latter trend is exemplified by the approaches to eco-fascism in the 1990s and the proposal by James Lovelock of the need for a benevolent dictator to take control of the Earth and tackle the global ecological crisis.[6] Some recent scholarship still advocates for a pragmatic environmental authoritarianism, pointing to the success of China's approach to environmental management as a model[7] as well as the experience from other nations like Korea[8] and Singapore.[9] Nevertheless, Mick Smith warns that authoritarian interventions may seem attractive as a short-term solution, but they pose the risk of permanent tyranny.[10]

James Scott identifies authoritarian politics beyond authoritarian regimes in the phenomenon he termed high modernism, which involves the standardization of social practices to increase their legibility to the state and facilitate territorial control, including control over the environment.[11] While Scott finds a univocal relation of expansive domination in technologies of government, Timothy Mitchell detects a more nuanced relation between human and non-human, as well as intentional and unintentional components in these technologies, emphasizing that despite the apparent existence of knowledge and knowledge objects as separate realms, they are indeed the result of a pragmatic co-construction.[12] The consequence of Mitchell's perspective on the politics of expertise for the study of authoritarianism is that, despite myths of rationality and control associated with authoritarian regimes, their practices of government are still frequently experimental and pragmatic.

There is a palpable affinity between Mitchell's understanding of knowledge politics and some recent scholarship devoted to the histories of the Cold War, particularly the detailed examinations of Cold War technopolitics[13] and environmental politics[14] that emphasize practices of government rather than the use of abstract ideologies as the primary explanatory factors for historical trajectories. Gabrielle Hecht's *Entangled Geographies* moves beyond hegemonic narratives of bloc politics to examine Cold War dynamics of knowledge and technology, a productive perspective for understanding the contingency of authoritarian regimes operating at the fringes of world politics. Laakkonen et al.[15] call for renewed attention to the environmental history of the Cold War, moving away from the assumption that ideological disagreements formed the kernel of this global conflict. Instead, they provide a nuanced understanding of the dynamics of natural resource exploitation and the emergence of nature as a global political concern, toward a better understanding of the dynamics of exploitation of natural resources, the emergence of nature as a global political concern and the proliferation and diffusion of environmentalisms across bloc divides.

The work of Stephen Brain is exemplary of the new trends in the study of authoritarian politics and the environment. He argues that during the Cold War superpowers were willing to accept environmentalism to improve the appeal of their ideologies even if it was not necessarily central to their political ideas,[16] a point that has been illustrated elsewhere with early environmentalist initiatives led by Stalin himself.[17] Fernando Estenssoro has pointed out that peripheral nations had fewer incentives to pursue environmental protection, describing the adoption of environmental regulations as primarily an external imposition.[18] This adoption was not always a top-down matter of technocratic politics and in some cases, it gave rise to new forms of political contestation, as shown by Viktor Pál with the case of Cold War Hungary, where the cynical uptake of environmentalism by a dictatorship ended up backfiring with the emergence of an environmental movement aligned with opposition to the regime.[19]

In Chile, Pinochet's dictatorship has traditionally been painted as unconcerned with environmental matters and environmentalism more general, analyzing environmental matters as the outcome of economic policy like the institutional approach used by Carl Bauer[20] and the political ecology of Rayén Quiroga.[21] This position has been useful to highlight the introduction of environmental regulations during the 1990s as progressive political innovations that marked a stark contrast with the authoritarian past. This view, however, has been challenged by authors like Mauricio Folchi[22] and Sebastián Ureta,[23] whose work emphasizes the existence of the environment as a matter of concern and subject of government for the dictatorship, stressing the continuities in the general directives of environmental policy between the dictatorship and later democratic governments. Such work has opened new research agendas that emphasize the examination of situated policy practices instead of monolithic ideological narratives, complicating historical narratives that automatically attach ideological stances to certain attitudes toward the environment, and giving more weight to the mediating role of geography and technology to the environmental outcomes of authoritarian regimes.

The coppering of Quintero Bay before the coup

In the 1950s, the Chilean government was following a major industrialization plan for developing heavy industrial facilities and increasing its power generation capacity, following a path initiated a decade earlier with the ascendance to power of the Radical Party and its economic policies centered on the entrepreneurial role of the state. This process was influenced by economic principles of import substitution and sovereignty over natural resources, important pieces of the Economic Commission for Latin America and the Caribbean's intellectual contribution. The industrialization plan had a strong focus on developing full processing capabilities in the copper sector—a strategic natural resource for Chile, which until then had been controlled almost entirely by foreign capital. In these circumstances, the construction of a new copper complex, which included a refinery, became crucial for the ambitions of the state and industry to control copper mining all the way from the point of extraction to the sale of refined copper bars.

In 1958, the National Foundries Enterprise (ENAF) approved the construction of a smelting and refining copper complex of Ventanas located in Quintero Bay on the central Chilean coast. The complex was only approved after a fierce competition between many towns that sought to secure for themselves the expected economic benefits of housing the copper complex, and a relatively comprehensive technical assessment that involved conducting the first environmental impact assessment for a Chilean infrastructure project.[24] The domestic enthusiasm regarding the construction of this vital piece of infrastructure for the copper sector was shared by international allies including the then German Minister for Economic Cooperation, Walter Scheel who, during a visit to the copper complex construction site in Ventanas, suggested that the project was evidence that "Chile is not an underdeveloped country but a nation in full swing toward development."[25]

Simultaneously, experts from the World Health Organization (WHO) expressed concerns regarding the air quality conditions in Chile, particularly the air quality of the city of Santiago, the SO_2 emissions already produced by the mining sector, and the threat posed by pollution from new infrastructure projects slated for construction in the coming decades. These conversations led to the Institute of Occupational Health and Air Pollution (IHTCA) opening its doors in 1963, a department in charge of among other tasks, overseeing environmental controls in Chile and providing training for government professionals from other Latin American countries. This was the first government organization with an explicit environmental regulation mission in Chile, and it was the first time that resources from a major international development institution, the United Nations Special Fund (predecessor of the UNDP) were used to fund an environmental initiative[26]—in other words, the first time in the world that money earmarked for economic development was approved for uses in environmental protection. Domestically, the IHTCA's portfolio was heavily influenced by the government's framing of the environment as a matter of public health.

In 1967, the IHTCA attempted to use its legal mandate to intervene in a conflict between the Ventanas copper complex and a group of Quintero Bay farmers affected by acid rain; this was the beginning of an environmental conflict that has lasted for more than half a century. The Ventanas copper complex, which included a coal-fired power plant in addition to the copper processing infrastructure, had started its operations in 1964, and despite the initial assurances from government experts, it caused environmental damage in the form of acid rain shortly after its opening. One of the reasons for the poor environmental performance of the complex was that several key components for mitigating air pollution, most notably including a sulfuric acid plant, had been pruned from the final construction plans to save costs, given the meager financial situation of Chile in the late 1950s. However, while the IHTCA could theoretically impose sanctions in cases where its regulations were violated, the vague and untested legal framework surrounding its authority hampered its ability to act.[27] Instead, members of Chile's national Parliament, the Ministry of Agriculture and local politicians from Quintero Bay insisted that the National Mining Enterprise (ENAMI), which operated the Ventanas copper complex, take the lead in solving the issue as the result of a round of negotiations. In 1969, ENAMI announced their decision to buy a sulfuric acid plant to fix Quintero Bay's air pollution problem, the purchase was proposed as part of a trade and technological cooperation agreement recently signed between Chile and Ceauşescu's Romania. That agreement was part of Ceauşescu's effort to appear as an independent economic actor in the global context after the Soviet invasion of Czechoslovakia, an effort well received by the Christian Democratic government of President Eduardo Frei Montalva.

However, after the election of President Salvador Allende in 1970, the cooperation agreement with Romania was suspended and the plan to import the acid plant was dropped.[28] Commercial relations with Romania had been frozen apparently due to the request of the Soviet Union to suspend bilateral trade negotiations with socialist countries, a condition readily accepted by Allende who expected a funding agreement for Chile like the one enjoyed by Cuba. Additionally, Chilean experts had realized that the technology used for smelting in Quintero Bay was outdated and inherently environmentally destructive, announcing just a few months before the military coup of September 11, 1973, that new smelting facilities, based on flash technology and including sulfuric acid processing facilities, would be constructed in cooperation with the Finnish corporation Outokumpu.[29]

The process of improving and expanding the copper sector's productive infrastructure was fundamental to making the industry more competitive and able to cope with the trade sanctions imposed after the nationalization of copper in 1971. Allende, a medical doctor by profession, expressed a general concern for environmental matters and formed the first commission tasked with providing a solution to air pollution in Santiago. However, air pollution in Quintero Bay and other mining areas received much less attention and was arguably seen as an unavoidable cost of ensuring the viability of the copper industry during times of great international pressure and trade uncertainty.

Isolation and dependency

On the September 11, 1973, Chilean democracy was overthrown by a violent coup led by Army General Augusto Pinochet, and coordinated as part of a military and civilian plot against President Salvador Allende's Popular Unity government. The startling violence of the coup itself, which resulted in the bombing of the presidential palace and Allende's death, was followed by a more prolonged wave of violence and human rights violations, including the imprisonment, torture, disappearance, and execution of thousands of perceived political opponents of the dictatorship. Pinochet's regime faced almost unanimous international condemnation and faced broken diplomatic relations as well as trade sanctions, some of which restricted Chile's access to more environmentally friendly mining technologies for the extensive copper operations controlled by the state-owned National Copper Corporation (CODELCO) and ENAMI. After abandoning initiatives that addressed existing environmental problems in the mining sector, the regime was forced to reconsider its prioritization of the environment in the 1980s when the threat of additional trade sanctions emerged as a response to the relative advantage of the Chilean copper sector over its U.S. competitors who had to comply with increasingly stringent environmental regulations.

An immediate consequence of the military coup of September 11, 1973 was the international condemnation of the move and the unilateral cancellation of diplomatic and trade relations by a large majority of governments. Chile's political and economic isolation restricted options for purchasing mining technologies capable of addressing the air pollution issues present in the sector. This need to secure technology was exacerbated by the fact that at the time of the military coup, several mining projects, including the expansion of production at Quintero Bay, were under evaluation. The investment in the Ventanas copper complex generated important dependencies for the Chilean government, which lacked proper funding to cover maintenance and upgrades.

Since 1968, there had been a clear acknowledgment that the technology used in Ventanas was insufficient to guarantee air quality in Quintero Bay, leading to a search for viable alternatives to provide a technological fix for the acid rain conditions of the area. As mentioned earlier, the first solution suggested was the purchase of a sulfuric acid plant from Romania as part of the ongoing trade agreements between both nations subscribed in 1969.

Romania was considered at that time an emerging actor in the development and commercialization of mining technologies and the agreement was attractive for Frei Montalva's government because it involved exchanging goods and services instead of currency that was very limited at the time. The overall agreement was not looked upon favorably by President Salvador Allende, who suspended most trade cooperation with Romania after assuming office in 1970, including the purchase of the sulfuric acid plant. Allende wanted a closer relationship with Moscow, and to that end decided to halt initiatives that could be perceived as threatening to the hegemony of the Soviet Union in the Eastern Bloc.

During Allende's administration, "flash" smelting technology sold by the Finnish corporation Outokumpu emerged as an effective alternative to deal with air pollution at Quintero Bay and at the same time deal with the technological challenges of increasing and stabilizing production. However, the Finnish government discontinued development assistance to Chile two days after the military coup.[30] Four months later, in January 1974, the Finnish government definitively cancelled ongoing credit agreements for the Chilean mining sector and Outokumpu cancelled the agreement for the flash-based, copper smelter planned for Quintero Bay.[31] Despite the ethical inspiration that led Finnish authorities to cancel trade cooperation with Pinochet's regime, the diplomatic move had the unintended consequence of preventing Chilean copper corporations from accessing cleaner copper processing technologies for more than a decade.

The need to overcome trade isolation during the early years of the military dictatorship led to unorthodox agreements with other regimes. An example of the initial unorthodoxy of the regime comes from Arturo Yovane's short period as Mining Minister between 1973 and 1974. Yovane used the mining post to seek international political support for the dictatorship in exchange for special trade deals, the peak of this process was a tour in 1974 that included visits to Zambia, Iran, Saudi Arabia, and Romania. In clear contrast with the principled position adopted by Finland, the Communist dictatorship of Romania, led by Ceauşescu, condemned the military coup led by Augusto Pinochet but, in an odd move, kept all the diplomatic and trade channels open for continuing cooperation between both nations. Romanian officials requested their Chilean counterparts to resume the trade agreements previously left at a standstill by Allende's government. For Romania, those agreements were relevant to secure raw materials from Chile, while for both countries, there was an assumption that political support was an important asset for a dictatorship.

During an official visit of General Yovane to Romania in June 1974, he signed a new agreement for trade cooperation between Chile and the eastern European nation, including joint copper mining ventures and infrastructure development. This agreement included the investment of US$65.6 million for a sulfuric acid and fertilizer plants for the Ventanas copper complex.[32] Both governments planned to get funding for this infrastructure using development loans obtained by Romania to get around potential international sanctions against Chile, making explicit the precariousness of the Chilean position at the time. Interestingly, only the development of joint mining ventures prospered after the agreement was resumed in 1974, the sulfuric acid plant project was once again put on hold because of funding difficulties added to the consolidation in power of neoliberal technocrats opposed to public investment, as will be seen in the following section of this chapter.

Another example of the challenges faced by the dictatorship in terms of international dependency was the need to secure funding for new power generation infrastructure to respond to the growing demand for electricity. By the beginning of 1974 the most urgent project was a 200 megawatt coal-fired power plant in Quintero Bay, called Ventanas II. The position of the Chilean government, represented by the Minister of Economic Coordination and Development Raúl Sáez, was that without an urgent investment in electricity infrastructure, the country would face serious power shortage by 1977. Sáez, visited the US in May 1974 and met John Hennessy, the Treasury's Assistant Secretary for International Affairs, and Walter Sauer, Vice President of the Ex-Im Bank. These meetings led to the approval the same year of a loan to build the new power plant in Quintero Bay and the negotiation of a new US financial policy regarding Chile and its political status.

Sáez, who had a long history working in collaboration with US institutions, achieved the normalization of financial relations between Chile and the US, which included support for further loans taken from international organizations like the World Bank and the Inter-American Development Bank that funded the replacement in 1977 of the stack at Ventanas smelter.[33] The loan for the power plant approved on September 11, 1974, one year after the coup, was a reminder of the priorities of both governments whose leaders conspired to overthrow a democratic government and after a very brief moral hiatus, made up again to keep investment running, despite the highly publicized human rights violations happening in Chile. Robert McNamara, director of the World Bank, justified his position of cooperation with dictatorships on the basis that he was mandated to take into consideration economic and not political factors,[34] implying that while the political situation of Chile was a bloody dictatorship, the affinity of economic ideologies sufficed to get the World Bank's support. The approval of the power plant and the new stack was a crucial moment for Quintero Bay and showed complete disregard for its residents, as those loans sealed the expansion of the Ventanas industrial park, adding in 1977 another source of air pollution to the area and expanding the production capacity of the existing copper smelter, without introducing any relevant environmental remediation mechanism.

The ambiguity of the United States toward Chile during Pinochet's dictatorship was exemplified by a condemnation of human rights violations in Chile, while at the same time Chile became its main supplier of copper in the early 1980s. However, that trade partnership in the copper sector was threatened by the pressure exerted by American members of Congress from copper producing states who pushed for the imposition of sanctions or duties on Chilean copper, based on unfair competitive advantages owing to the lack of environmental regulations in the Chilean copper sector.[35] The introduction in Congress of the Copper Environmental Equalization Act of 1985 by Senator DeConcini was the most tangible expression of this attitude which, in turn, led to environmental improvements in the Chilean copper sector, with the specific announcement during the same year of the purchase of a sulfuric acid plant for the Ventanas copper complex. US support helped the Chilean dictatorship navigate the complications of international sanctions and isolation. Interestingly, instead of human rights condemnations, the most effective mechanism to improve life conditions in Quintero Bay was the threat of trade sanctions.

Ideological intransigence

The influence of isolation and economic dependency only partially explains the limited response of the Pinochet regime to environmental issues and particularly to air pollution matters in Quintero Bay. The influence of neoliberal technocrats also played an important role in minimizing attention to these issues due to their ideological re-conceptualization of the environment as an exclusively economic problem. Although neoliberalism is often depicted as an ideology imported to Chile that serves as an explanation for the reform processes undertaken by Pinochet's dictatorship, it nevertheless is important to point to the elective affinity between the regime's need for sources of rationality and legitimacy and the theoretical coherence and international prestige offered at the time by neoliberalism. Overall, Pinochet's dictatorship failed to solve the most pressing environmental problems faced by Chile in the 1970s and 1980s. However, it was very successful in establishing neoliberalism as the logic for future Chilean environmental policy.

The technocrats who pioneered neoliberal environmental policy—and indeed, many other kinds of neoliberal policies—in Chile were part of a civilian faction known as the "Chicago Boys" who worked closely with the dictatorship.[36] The Chicago Boys were Chilean economists trained at the University of Chicago under the supervision of prominent figures of monetarism including Milton Friedman and Arnold Harberger, thanks to a development agreement co-signed by the prestigious Pontifical Catholic University of Chile in the 1950s. Some of these economists collaborated in the sabotage of Allende's government and many were appointed to government positions by Pinochet from the beginning of his rule. The National Planning Office (ODEPLAN) soon became the stronghold from which the Chicago Boys wielded significant power over the authorization, design, and planning of government-funded infrastructure projects.

In essence, ODEPLAN's openly neoliberal agenda was to re-order the relation between the state and the economy through the deregulation of economic activities and the application of the subsidiarity principle that favored private solutions for public problems, and state intervention only in qualified cases where the private sector was unable to deliver solutions. ODEPLAN imposed strict restrictions on public investment to enforce the subsidiarity principle, demanding high social discount rates for investments. In addition, state-owned corporations were required to end their dependence on government subsidies and become self-reliant, dedicating their available funds to increase short-term profitability, rather than investing in longer-term improvement in efficiency or environmental performance. Although the latter was combined with a resistance to impose any regulation that would force corporations to make investments that could hinder profits, ODEPLAN was wary that environmental matters opened a particularly vulnerable flank in terms of the need for state intervention:

> The expanding industrialisation and the application of technologies that seek a larger production are provoking a serious damage in the atmosphere, the soil, the rivers, and the seas, a situation that is seriously worrying not only in highly developed nations, but also in those who are striving for a vigorous industrial take off ... In a social market economy like ours, it is necessary to establish criteria and norms to preserve, within the regions of the country, the physical and social well-being necessary for the survival of the human being. However, it is necessary to avoid inadequate norms that could introduce additional obstacles to the investments required for development.[37]

These developmentalist policy guidelines established by ODEPLAN in the early years of the regime exemplify its recognition of environmental problems as an important matter, while at the same time posing regulatory standards as a danger to the country's economic performance. ODEPLAN Chicago Boys' position against public investment and the adoption of environmental regulations that could undermine economic development limited ENAMI and other government departments' ability to address, much less resolve, the environmental crisis of Quintero Bay.

Shortly after the establishment of the dictatorship, a group of farmers from Quintero Bay contacted the Ministry of Agriculture to express their support for the new government and to ask for their long-standing issues with air pollution to be addressed, specifically through compensation for their financial losses and the construction of a sulfuric acid plant.[38] As evidence of their claim, the farmers attached documents from a 1967 inspection conducted by experts from the Ministry of Agriculture. They also publicized their case through a story in the newspaper *El Mercurio de Valparaíso* in August 1974, interestingly this story was not censored, probably because environmental problems were not yet perceived as political ones. The

Agriculture Minister even sympathized with the farmers and sent a four-point proposal asking the Ministry of Mining and ENAMI to take action regarding the pollution:

1 Lifetime compensation payments from the budget of ENAMI
2 Expropriation and relocation to an area more benign to life
3 Payments for property damage
4 Anything that could grant a compensation to farmers for all these years of losses.[39]

The Vice President of ENAMI readily acknowledged the pollution issue and assured farmers that their state corporation, overseen by the Ministry of Mining, was seeking a solution:

> Since Ventanas smelter was set up in 1964, it has been emitting through its stack approximately 50,000 Nm3/hr [normal cubic meters per hour] of gases with 0.5 to 1% of SO2 [sulfur dioxide], from the reverberatory furnace and 100,000 to 150,000 Nm3/hr with approximately 4% of SO2 corresponding to the converters. Both figures will increase in 20% in the future as a result of the expansion program [...] ENAMI has been concerned with the development of a project to install a sulphuric acid plant that operates using conversion gases and at the same time it is considering to adopt modern processes that prevent the emission of SO2 in any future expansion.[40]

At that time, ENAMI proposed two possible solutions for the rampant pollution in Quintero Bay: either, the already mentioned, import of a sulfuric acid plant from Romania, ratified by Minister Yovane in 1974; or the construction and operation of a plant through a consortium of five Chilean chemical companies. The latter plan had strong support from the Treasury Minister. Both proposals were rejected by ODEPLAN, even though the profitability of running a sulfuric acid plant in Quintero Bay was above their own 18 percent minimum for public investments. Instead, ODEPLAN's position was bluntly ideological: ODEPLAN argued that there was no reason for the state to invest, since a private company should be perfectly capable of constructing and operating a plant without state assistance.[41] ODEPLAN maintained its commitment to this principle of subsidiarity, even despite the urgent need for a solution for the residents of Quintero Bay affected by air pollution, as well as for the mining sector in central Chile that had been experiencing a shortage of sulfuric acid for several years that could have been fixed with the installation of a plant at the Ventanas copper complex. The scarcity became so severe that SONAMI, the private mining industry association, sent a letter to the Ministry of Mining asking for the construction of a US$14 million sulfuric acid plant in Quintero Bay that would increase the availability of acid for the mining sector and provide an "ecological solution to prevent the pollution produced in the proximity of Ventanas that has deteriorated its agriculture and the health of its inhabitants."[42]

Nonetheless, ODEPLAN continued to prevent ENAMI from purchasing the equipment necessary for environmental remediation in Quintero Bay. Instead, in August 1977 ODEPLAN proposed the experimental idea of holding a public tender for the fumes emanating from Quintero Bay's smokestacks, which could be converted into sulfuric acid if the purchaser constructed a plant.[43] Despite advertising the project internationally, it did not receive even a single bid. The Chilean Association of Chemical Industrials (ASIQUIM) sent a letter to ODEPLAN complaining about the terms of the tender, specifically regarding the transfer of responsibility for all ecological consequences of air pollution from ENAMI to the purchasers of the fumes.[44] According to ASIQUIM, converting the fumes into sulfuric acid would not solve the totality of the environmental crisis of Quintero Bay, as was apparently suggested by ODEPLAN's call.

The Ministry of Mining persisted in seeking funding for the sulfuric acid plant, including funds for its construction in its 1981 budget. However, ODEPLAN rejected the investment again. Sergio Melnick, head of ODEPLAN's Investment Department, summarized their position in this matter:

> [W]e are opposed to state enterprises undertaking or increasing investments in areas that can be perfectly developed by the private sector … As a general principle, ODEPLAN opposes that public enterprises increase their size or make investments that end up increasing their participation in the productive apparatus.[45]

At this point, the environmental crisis of Quintero Bay reached the highest levels of Pinochet's government, the Junta, which had been unaware of the controversy between ODEPLAN and the Ministry of Mining. In 1981, Captain Herve Dilhan, Undersecretary of Mining, explained to the Junta that:

> Cleaning ENAMI's fumes would have been ready, but we had a problem of differing criteria with ODEPLAN, which decided that this project was not carried out by ENAMI and fumes should be tendered. We are bidding for the last time. If sold [fumes] to produce sulphuric acid, the problem will be solved. Otherwise, ENAMI will have to invest fourteen million dollars in the plant.[46]

The intransigence of ODEPLAN was finally overcome in 1985 by pressure from international trade partners concerned about the environmental performance of the Chilean copper sector, as mentioned in the preceding section, leading to the announcement that a sulfuric acid plant would be built. It can be argued that ODEPLAN was more concerned with setting a precedent about how environmental problems should be dealt with in

Chile—by the market rather than the state—than providing readily available, effective and even profitable solutions for existing problems. In this sense, Quintero Bay became an experimental site for these technocrats, who were willing to sacrifice human lives and the economic profits of several corporations in the name of their philosophical stance.[47] ODEPLAN promoted the need for market mechanisms that at the time mostly existed only in the theoretical imaginations of neoliberal economists. This was also the approach taken to solve the air pollution issue in Santiago: instead of applying direct government action and regulations that had been successful elsewhere, ODEPLAN delayed acting by sending officers to study the feasibility of cap and trade emissions schemes in Texas in the early 1980s.[48]

Constitutionalization of the environment

Pinochet's dictatorship shaped the fate of environmental problems in Chile applying a strong neoliberal agenda, outsourcing state responsibilities for environmental matters to the private sector through the creation of market mechanisms. This process included an additional step that was adopted in parallel to the debates about the construction of a sulfuric acid plant in Quintero Bay, namely, the "constitutionalization" of the environment. The environment was introduced as a legal entity recognized at a constitutional level in the new political constitution of 1980. This definition of the environment was fundamental in establishing the legal and political conditions for neoliberalism to operate, setting the boundaries for the meaning of the environment and environmental responsibility in Chile.

Pinochet's dictatorship promoted the narrative that the coup that deposed President Allende was justified on the basis that his government was leading the country to a situation of chaos with no respect for the rule of law. For Pinochet's regime, Allende's government was a symptom of a deeper problem contained in the constitution of 1925, which granted ample social rights to the point of becoming what Jaime Guzmán, a prominent legal adviser to Pinochet, called a "totalitarian democracy."[49] Shortly after taking power, in September 1973, the junta announced a commission to study a new constitution, La Comisión de Estudios de la Nueva Constitución Política de la República de Chile (CENC), also known as the Ortúzar Commission. The stated aim of the commission, which was staffed mostly by conservative legal experts, was to produce a document that could safeguard democracy in the future from unreasonable social demands. A new constitution was approved in a fraudulent plebiscite on September 11, 1980 and enacted on March 11, 1981. One of the most prominent features of this new constitution was the strict delimitation of twenty-six constitutional rights, including Article 19 Section 8: the right to live in a contamination-free environment.

The constitutional right to live in an environment free of contamination was a legal innovation, one that could be seen as very progressive from both a historical and global perspective, especially because the constitution explicitly allowed for other rights, specially property rights, to be infringed in the name of environmental protection. However, there were severe limitations to the right to a pollution-free environment. The constitution itself was extremely vague on how issues of the environment and pollution should be defined and enforced. CENC proceedings, which have legal standing in the interpretation of constitutional law, suggested that the state would establish objective, scientific thresholds to determine acceptable levels of pollutants—a process that was not explicitly guaranteed to protect human health. Moreover, the CENC suggested that thresholds be incrementally reduced to their intended levels over time, enforcing a hierarchy that placed national economic development above the right to live in a pollution-free environment. Finally, the recourse to pollution thresholds as the sole means of environmental protection reflected the CENC's baseline idea of the environment as always already polluted, a conception that de-emphasized the role of industrial pollution in causing environmental damage. Industry was further protected from environmental responsibility by the fact that the regime never established a single threshold regarding pollutants.

Farmers from Quintero Bay, who had been affected by the Ventanas copper complex for over a decade and let down several times by the government, were the first to attempt to exercise their constitutional right to live in a pollution-free environment with the submission of two "constitutional remedy" applications in 1981. Both applications were rejected by the courts, with three main arguments delineated in two successive decisions: the lack of evidence that fumes had caused agricultural losses, the existence of previous litigation between the farmers and ENAMI, and finally, the technical objection that the source of pollution was neither illegal nor "arbitrary" under Chilean law. This ruling angered a well-known conservative legal scholar who published an article commenting on the absurdity of these decisions that privileged legal formalities instead of protecting the lives and pecuniary interests of those affected by pollution.[50]

The news of the constitutional remedy case presented by Quintero Bay farmers concerned junta members. Air Force General Fernando Matthei, who took a special interest in the issue, since he had a house in Maitencilo near Quintero Bay. claimed:

> I believe that we are being too lenient with miners and we are not applying the same standard for other citizens.[51]

The junta asked the Ministry of Mining to find a solution, despite the court's ruling denying the farmers' claims against ENAMI. The junta's sudden interest in the case took place against a complex backdrop of ongoing efforts from multiple government departments to address conflicts between mining and agriculture. In 1980, a commission had been formed with participation from the Ministries of Mining, Health, and Agriculture as well as ODEPLAN to propose permissible levels of contaminants affecting both human populations and ecology in areas with mining activity.[52]

The commission identified water and air pollution as the most important health concerns and dam tailings and air pollution as the most pressing agricultural issues, pointing to forty-four mining sites as critical hotspots. However, in the end the commission refrained from proposing contaminant thresholds based on three arguments: first, the commission considered available scientific data insufficient for establishing restrictions on the emission of pollutants; second, the commission decided that the right balance between development and environmental protection was already in place, since most mining operations were located in remote and unpopulated areas such as deserts; third, the commission warned that "anti-pollution" policies had forced the closure of copper smelters in the United States and could do the same in Chile.[53] In sum, the commission concluded that the "social benefits" of copper justified some concessions, particularly in trading the protection of human health for the achievement development goals. They declared themselves unable to determine permissible levels of pollutants until the government defined the appropriate trade-off between health and environmental protection on the one hand and economic development on the other.

Another governmental commission soon followed the same path. In 1984, the Ministry of Agriculture asked the National Standards Institute (INN) to create a commission to establish environmental quality standards for levels of air and water pollution in areas of agricultural and forestry activity or wildlife presence. This commission gathered participants from various public sector departments as well as representatives from academia and industry. Its mission was framed as technical, but the commission failed to reach any conclusions regarding standards. The INN later dissolved the commission on the basis that, instead of a technical matter, it was being asked to intervene in a political conflict between clashing perspectives about national development.[54]

At the same time, in 1984, the Ministry of National Assets created the National Commission of Ecology (CONADE), to provide policy advice to the President of the Republic in matters of environmental protection and the conservation of natural resources.[55] CONADE, through a legislative commission, prepared a draft for a General Environmental Protection Law. The draft was widely rejected by private and public sector actors, particularly the mining sector, which had been excluded from participating in the drafting process, likely because of their obstructionist role in previous initiatives opposing any environmental regulation for the mining industry. The mining sector presented two main arguments for rejecting the proposal: first, they claimed the law lacked proper understanding of the technical aspects of mining facilities and the ecology of their surroundings, and second, that the law failed to achieve balance between technical-economic concerns and the environment.[56]

During the final years of the dictatorship, the Ministry of Mining organized an advisory committee for environmental matters in response to increasing pressure from communities affected by mining pollution and the threat of international trade sanctions. The committee was made up of representatives from professional and business associations as well as state-owned mining corporations. The committee released a document that emphasized the need for balance between environmental protection and what they perceived as the necessary use of natural resources for social and economic development, claiming that only new facilities should be required to comply with environmental regulations. The commission recommended clear, realist, and flexible environmental law that would not interfere with the current business model of the mining sector—effectively stating that environmental regulations should adapt to the mining sector instead of the other way around.[57]

The idea of creating scientific standards to define pollution permeated outside government special commissions and prompted a wave of research aimed at proving the existence of environmental pollution. Arguably, the introduction of the right to live in a pollution-free environment motivated the work of these scientists who sought to provide evidence to support the cases of communities affected by toxicity. One of the most well-known cases concerns the pollution in Quintero Bay, where Sergio González and his team from the Institute for Agricultural Research (INIA) conducted the first scientific studies proving the extensive impact of mining on soil and animal health.[58] These findings were later supported by veterinarian Eduardo Schmidt Funes, who moved to Quintero to establish a pig farm in the late 1980s and shared results from soil and water tests at a Rotary Club conference in 1989.[59] Finally, Jaime Chiang, a chemical engineering PhD and former Rector of the Federico Santa María University, led a team of students from the University of Valparaíso in conducting soil studies outside the Ventanas copper complex and testing ENAMI workers to demonstrate the severity of their exposure to heavy metals.[60]

Despite these and other attempts by citizens, scientists, and government officials from several ministries to hold the mining industry accountable for its environmental impacts using available institutional mechanisms, they were confronted with its immense political power inside and outside government. The constitutional right to live in a contamination-free environment became for the most part an aspirational rather than material right, because defining legal environmental thresholds against which to measure claims of contamination became an impossible task. The impossibility of the task was largely due to the unwillingness of the mining sector to accept any restriction to their activities.

Conclusions

More than fifty years have passed since the opening of the highly polluting Ventanas copper complex in Quintero Bay in 1964, but it remains today one of the most toxic areas in Chile. The industrial park around the copper complex has expanded to more than twenty different polluting and dangerous industries and frequently makes headlines in Chilean media for new toxic incidents, sparking familiar cycles of outrage, protracted legal battles and virulent political conflict. The mechanisms that have created the vicious, chronic conditions of Quintero Bay's environmental problems can be traced back to Pinochet's dictatorship.

This chapter contributed to a more general understanding of the environmental policies and practices of authoritarian regimes. The account presented here, with its focus on Quintero Bay in Chile, is relevant to the trajectory of environmental politics in right-wing authoritarian regimes in Latin America as well as to the impact of trade, technology, neoliberal orthodoxy, and environmental constitutionalism to the production of chronic pollution in specific sites. Pinochet's regime interrupted the nascent institutionalization of the environment as a matter of public health by previous governments and abruptly shifted toward policies that sought to materialize the ideological abstractions of neoliberal technocrats and the de facto subordination of human health to economic development goals. Many of these ideological trends outlasted the dictatorship as part of the environmental legislation passed since the return to democracy.

A broad agenda for research can be outlined from the issues examined in this chapter: first, the epistemological transitions experienced by the institutions like the IHTCA after the military coup of 1973, specifically how ideological and technological shifts reshaped air pollution from a matter of public health to one of economic policy; second, the connection between internal political pressure and international relations, in terms of trade and technological development, and their influence into the regulatory responses that the Chilean state has historically provided for air pollution matter; third, the outcomes of existing environmental laws and policies in Chile in terms of environmental justice, considering the persistence of air pollution as a chronic issue with limited prospects of improvement anytime soon, despite the existence of environmental regulations.

Notes

1 Renato Cristi, *Carl Schmitt and Authoritarian Liberalism: Strong State, Free Economy*, Political Philosophy Now (Cardiff: University of Wales Press, 1998).

2 Mauricio Folchi, "La Gestación de la Política Ambiental Minera Antes de la 'Revolución Ambiental de los Noventa,'" in E. Aliste and A. Urquiza (eds.), *Medio ambiente y sociedad: Conceptos, metodologías y experiencias desde las ciencias sociales y humanas* (Santiago, Chile: Ril Editores, 2012); Sebastián Ureta, "'Because in Chile [Carbon] Markets Work!' Exploring an Experimental

Implementation of an Emissions Trading Scheme to Deal with Industrial Air Pollution in Santiago," *Economy and Society* 43, no. 2 (2014): 285–306.

3 Carolina Riveros, "El Problema de La Contaminación Atmosférica En Santiago de Chile: 1960–1972" (Santiago, 1997).

4 Pablo Ulricksen et al., "Aire," in N. Gligo (ed.), *Informe País: Estado del Medio Ambiente en Chile* (Santiago, Chile: INAP, Universidad de Chile, 2016), 39–114.

5 Murray Feshbach and Alfred Friendly, *Ecocide in the USSR: Health and Nature under Siege* (New York: Basic Books, 1992); Barbara Jancar-Webster, *Environmental Management in the Soviet Union and Yugoslavia: Structure and Regulation in Federal Communist States*, Duke Press Policy Studies (Durham, NC: Duke University Press, 1987); Boris Komarov, *The Destruction of Nature in the Soviet Union* (London: Pluto Press, 1978); Charles E. Ziegler, *Environmental Policy in the USSR* (Amherst, MA: University of Massachusetts Press, 1987).

6 James Lovelock, *The Revenge of Gaia: Earth's Climate in Crisis and the Fate of Humanity* (New York: Basic Books, 2007).

7 Mark Beeson, "Environmental Authoritarianism and China," in T. Gabrielson et al. (eds.), *The Oxford Handbook of Environmental Political Theory* (Oxford: Oxford University Press, 1st ed., 2016); Mark Beeson, "The Coming of Environmental Authoritarianism," *Environmental Politics* 19, no. 2 (March 2010): 276–294, https://doi.org/10.1080/09644010903576918.

8 Heejin Han, "Authoritarian Environmentalism under Democracy: Korea's River Restoration Project," *Environmental Politics* 24, no. 5 (September 3, 2015): 810–829, https://doi.org/10.1080/09644016.2015.1051324.

9 Heejin Han, "Singapore, a Garden City: Authoritarian Environmentalism in a Developmental State," *The Journal of Environment & Development* 26, no. 1 (March 2017): 3–24, https://doi.org/10.1177/1070496516677365.

10 Mick Smith, "Against Ecological Sovereignty: Agamben, Politics and Globalisation," *Environmental Politics* 18, no. 1 (February 2009): 99–116, https://doi.org/10.1080/09644010802624843.

11 James C. Scott, *Seeing Like a State: How Certain Schemes to Improve the Human Condition Have Failed*, Yale Agrarian Studies (New Haven, CT: Yale University Press, 2008).

12 Timothy Mitchell, *Rule of Experts: Egypt, Techno-Politics, Modernity* (Berkeley, CA: University of California Press, 2002).

13 Gabrielle Hecht (ed.), *Entangled Geographies: Empire and Technopolitics in the Global Cold War*, Inside Technology (Cambridge, MA: MIT Press, 2011).

14 J. R. McNeill and Corinna R. Unger, *Environmental Histories of the Cold War* (Cambridge: Cambridge University Press, 2010), http://public.eblib.com/choice/publicfullrecord.aspx?p=803042; Simo Laakkonen, Viktor Pál, and Richard Tucker, "The Cold War and Environmental History: Complementary Fields," *Cold War History* 16, no. 4 (October 2016): 377–394, https://doi.org/10.1080/14682745.2016.1248544.

15 Laakkonen, Pál, and Tucker, "The Cold War and Environmental History."

16 Stephen Brain, "The Appeal of Appearing Green: Soviet–American Ideological Competition and Cold War Environmental Diplomacy," *Cold War History* 16, no. 4 (October 2016): 443–462, https://doi.org/10.1080/14682745.2014.971015.

17 Stephen Brain, "Stalin's Environmentalism," *The Russian Review* 69, no. 1 (2010): 93–118.

18 Fernando Estenssoro Saavedra, *Historia del debate ambiental en la política mundial, 1945–1992: la perspectiva latinoamericana* (Santiago de Chile: IDEA Inst. de Estudios Avanzados, University de Santiago de Chile, 1st ed., 2014).

19 Viktor Pál, *Technology and the Environment in State-Socialist Hungary* (New York: Springer Berlin Heidelberg, 2017).

20 Carl J. Bauer, *Against the Current: Privatization, Water Markets, and the State in Chile, Natural Resource Management and Policy* (Boston, MA: Kluwer Academic Publishers, 1998).
21 Rayén Quiroga (ed.), *El Tigre Sin Selva: Consecuencias Ambientales de La Transformación Económica de Chile, 1974–1993* (Santiago, Chile: Instituto de Ecología Política, Area de Economía Ecológica, 1994).
22 Folchi, "La Gestación de la Política Ambiental Minera Antes de la 'Revolución Ambiental de los Noventa.'"
23 Ureta, "'Because in Chile [Carbon] Markets Work!' Exploring an Experimental Implementation of an Emissions Trading Scheme to Deal with Industrial Air Pollution in Santiago."
24 Leonardo Valenzuela Pérez, "Coppered Lives: The Chilean Sacrifice Zone of Quintero Bay" (The University of Sydney, 2016), http://hdl.handle.net/2123/15914.
25 SONAMI, "Fundición y Refinería En Ventanas," *Boletín Minero*, no. 667 (1962): 4268.
26 Riveros, "El Problema de La Contaminación Atmosférica En Santiago de Chile: 1960–1972."
27 Riveros.
28 Francisco Sabatini, Francisco Mena, and Patricio Vergara, "Otra Vuelta a La Espiral: El Conflicto Ambiental de Puchuncaví Bajo Democracia," *Ambiente y Desarrollo* XII, no. 4 (1996): 30–40. These authors incorrectly suggested that the sulfuric acid plant was bought by Allende's government from Romania and it was not delivered because of the military coup.
29 Joselín González, "Horno Flash Para Fundición de Cobre de 'Las Ventanas,'" *Boletín Minero* no. 702 (1973): 17–19.
30 EILEN, "Year 1973 in Finnish Foreign Policy," Text, Archive and Chronology of Finnish Foreign Policy, 2012, www.eilen.fi/en/chronology/1973/.
31 EILEN, "Year 1974 in Finnish Foreign Policy," Text, Archive and Chronology of Finnish Foreign Policy, 2012, www.eilen.fi/en/chronology/1974/.
32 US Department of State, "Joint Government Investment Agreement: Chile–Romania," Wikileaks, 1974, Public Library of US Diplomacy, https://search.wikileaks.org/plusd/cables/1974SANTIA03551_b.html.
33 Ministerio de Minería, "Providencias Con Antecedentes, Providencia 502," 1976, Archivo Nacional de la Administración, Chile.
34 Ann Crittenden, "Loans from Abroad Flow to Chile's Rightist Junta," *New York Times*, February 20, 1976.
35 Richard B. Dagen and Michael S. Knoll, "Duties to Offset Competitive Advantages," *Maryland Journal of International Law* 10, no. 2 (1986): 273–293.
36 Juan Gabriel Valdés, *Pinochet's Economists: The Chicago School in Chile* (Cambridge: Cambridge University Press, 2008).
37 ODEPLAN, "Estrategia Nacional de Desarrollo Regional 1976–1990," ODEPLAN (1975), 125–126.
38 Ministerio de Minería, "Providencias Con Antecedentes, Providencia 699," 1974, Archivo Nacional de la Administración, Chile.
39 Ministerio de Minería, "Providencias Con Antecedentes, Providencia 70," 1976, Archivo Nacional de la Administración, Chile.
40 Ministerio de Minería, "Providencias Con Antecedentes, Providencia 43," 1976, Archivo Nacional de la Administración, Chile.
41 "Posición 'Privatizante' versus 'Estatista' En Ácido Sulfúrico," *Minería Chilena*, 1981.
42 SONAMI, "Necesidad Mayor Producción Ácido Sulfúrico," *Boletín Informativo de La Sociedad Nacional de Minería* no. 35 (1978): 8.
43 "Posición 'Privatizante' versus 'Estatista' En Ácido Sulfúrico."

44 "Posición 'Privatizante' versus 'Estatista' En Ácido Sulfúrico."
45 "Por Principio Nos Oponemos a Aumentar La Inversión Del Estado," *Minería Chilena*, 1981, 17.
46 Junta de Gobierno, "Actas de Sesiones de La Junta de Gobierno, Acta 33," 1981, 128, Biblioteca del Congreso Nacional de Chile.
47 Andre Gunder Frank, "Economic Genocide in Chile: Open Letter to Milton Friedman and Arnold Harberger," *Economic and Political Weekly* 11, no. 24 (1976): 880–888. This author called that attitude regarding social policy "economic genocide."
48 Ureta, "'Because in Chile [Carbon] Markets Work!' Exploring an Experimental Implementation of an Emissions Trading Scheme to Deal with Industrial Air Pollution in Santiago."
49 Jaime Guzmán, "El Miedo: Síntoma de La Realidad Político-Social Chilena," *Portada*, 1969.
50 Eduardo Soto Kloss, "El Recurso de Protección y El Derecho a Vivir En Un Ambiente Libre de Contaminación," *Revista de Derecho y Jurisprudencia* 78, no. 3 (1981): 113–118.
51 Junta de Gobierno, "Actas de Sesiones de La Junta de Gobierno, Acta 33," 27.
52 Folchi, "La Gestación de la Política Ambiental Minera Antes de la 'Revolución Ambiental de los Noventa.'"
53 Folchi.
54 Folchi.
55 Pablo Camus and Ernst R. Hajek, *Historia Ambiental de Chile* (Santiago, Chile: Departamento de Ecología, Facultad de Ciencias Biológicas, Pontificia Universidad Católica de Chile, 1998); Folchi, "La Gestación de la Política Ambiental Minera Antes de la 'Revolución Ambiental de los Noventa'"; Mauricio Folchi, "La Insustentabilidad Del 'Boom' Minero Chileno: Cobre, Política y Medio Ambiente 1983–2003," *Ecología Política* no. 26 (2003): 23–50; Quiroga, *El Tigre Sin Selva*.
56 Folchi, "La Gestación de la Política Ambiental Minera Antes de la 'Revolución Ambiental de los Noventa.'"
57 Folchi.
58 Sergio González, "Acumulación Metálica En Suelos Del Área Bajo Influencia de Las Chimeneas Industriales de Ventanas (Provincia de Valparaíso, V Región)," *Agricultura Técnica* 50, no. 2 (1992): 214–219; Sergio González, "El Impacto de Fundiciones de Minerales Sobre Las Actividades Agropecuarias," *IPA La Platina* no. 39 (1987): 46–48; Sergio González and Enrique Bergqvist, "Evidencias de Contaminación Con Metales Pesados En Un Sector Del Secano Costero de La Quinta Región," *Agricultura Técnica* 46, no. 3 (1986): 299–306.
59 Eduardo Schmidt, "El Medio Ambiente En Medicina Veterinaria," *Monografías de Medicina Veterinaria* 12, no. 2 (1990); Eduardo Schmidt, "Los Efectos Económicos y Sociales Producidos Por La Contaminación Ambiental En Las Zonas Agrícolas," Conferencia Rotary International (Antofagasta, 1989); Juan Miguel Gastó, Eduardo Schmidt, and Mario Trivelli, *Medio Ambiente: ¿realidad o Moda?*, Documentos de Trabajo 143 (Centro de Estudios Públicos, 1990).
60 Jaime Chiang, "Desarrollo Industrial y Contaminación," *Ambiente y Desarrollo* V, no. 2 (1989): 43–50; Jaime Chiang, "Niveles de Metales Pesados En Organismos, Agua y Sedimentos Marinos Recolectados En La V Región de Chile," in *Revista Pacífico Sur* (Simposio Internacional de los Recursos Vivos y las Pesquerías en el Pacífico Sudeste, Viña del Mar: Comisión Permanente del Pacífico Sur (CPPS), 1989), 205–215; Jaime Chiang et al., "Determinación de Arsénico En Individuos Expuestos a Altos Niveles de Contaminación," *Revista Chilena de Nutrición* 18, no. 1 (1990): 39–50; Jaime Chiang et al., "Determinación de Cobre, Cadmio, Manganeso, Plomo, Zinc y Arsénico En Sedimento Atmosférico En La Zona de Quintero," *Boletín de La Sociedad Chilena de Química* 30, no. 3 (1985): 139–158.

6 Bark beetles and ultra-right nationalist outbreaks

Białowieża, Poland

Eunice Blavascunas and Agata Agnieszka Konczal

The status of the Polish Białowieża Forest, touted as Europe's last primeval forest, seems at times highly ambiguous. The forest is both logged and protected. It straddles the border with Belarus marking the eastern boundary of the European Union. It includes plantations and old growth. Since the fifteenth century, the forest has been a coveted symbolic place for various political regimes and occupying powers, including the Polish-Lithuanian Commonwealth, Czarist Russia, Germany in both world wars, the Soviet Union, and Poland in successive twentieth-century Republics.[1] The ambiguity of this forest has proven valuable to many incarnations of the state, which have mythologized the forest for political ends. Given the number of geopolitical changes, the forest is always purifying and in need of purification, either through hunting, logging, or preservation. For much of the twentieth century, the Białowieża Forest was not a centerpiece of Polish national heritage, but it has become so today under an ultra-nationalist government with authoritarian tendencies, the Law and Justice Party (PiS). PiS has taken to heavily logging the forest as its way to make the forest and the foresters embody Polish national virtues of martyrdom and sacrifice.

The current PiS administration (2015–2018) initiated a threefold increase in logging over 2017–2018 due to a bark beetle outbreak. Foresters believed that the beetle would have destroyed the forest had foresters not acted decisively against nature conservationists who felt that the beetle was renaturalizing the logged and managed part of the forest. The forest is divided between the Białowieża National Park (10,500 hectares), with almost no logging, and commercial forestry lands, operated by Polish State Forests (47,500 hectares). The EU sees this forest as part of the greater primeval forest that once covered Europe, whereas the current PiS administration sees the forest as a working Polish heritage. The PiS administration interpreted a popular mandate to rule as a license to flout constitutional rules and EU procedures in relation to Białowieża, including violating the EU's NATURA 2000 protection policies. The Białowieża Forest thus became one of the pivot points for European rule of law. We examine this multifaceted conservation conflict in light of claims that PiS has authoritarian-leaning tendencies. An outbreak of bark beetles provided the pretext for politicized logging at state properties adjoining the national park. But these insects of political convenience can be imagined/narrated differently, and they have an agency of their own beyond human meaning. Bark beetles may provide a new, better way of imagining the Białowieża forest that also helps us imagine ways to deflect authoritarianisms.

While past authoritarian rulers and occupiers of the forest might have constructed the mythologies behind their rule or occupation on charismatic game animals, or the oldest trees (Schama 1995; Daszkiewicz et al. 2012), this chapter turns to the native spruce bark beetle, *Ips typographus*. Through this insect we tell a story about the revival of a Polish nationalist myth used to fix the forest's biological composition and meaning. In this myth, the history of the nation is equivalent to the history of Polish forestry and the forest turns into the substitute of Polish territory, always being threatened by foreign powers. The forest, equal to the nation, has to be protected against foreign influences (Konczal 2018). The beetle is not treated as a Polish native but as a pretext for ridding the forest of pests that might allow foreign invasion. Extracting timber in Białowieża, because of the bark beetle, becomes the way to enact this myth. PiS used long-standing agents of the Polish state, foresters, as their central symbolic element of what it means to belong to the Polish nation.

Deciding if the current government is in fact authoritarian, or whether authoritarian rule has better protected or exploited the forest, is not our aim. Rather, we use an anthropological lens to show how biological changes to the forest's composition coincide with an equally fraught political transformation. In the tradition of environmental history and environmental anthropology we graft narratives of political change onto environmental change. Looking at how PiS displaces many stories of the forest (its various ethnic peoples, the forest's biological composition) through one controlling narrative, enacted against the democratic norms of the European Union, we are able to expose how narrative works in environmental outcomes where democratic norms are in question.

Poland's current PiS government (also in power 2005–2007 but with a more conciliatory position on the Białowieża conflict) used the beetle as a pretext for decisively challenging conservationist's interpretations of State Forest holdings as primeval. This motion came after a 2011 forestry management plan, designed with conservationists, to preserve the commercial forest's oldest stands from logging. While the Białowieża National park had been on the UNESCO World Heritage site list since 1979, the state forest acquired the status on the UNESCO World Heritage list in 2014, notably after the 2011 forestry management plan scaled back logging. Logging began in earnest in the summer of 2017 and the PiS government stated that logging was the only way to protect this forest and to keep it valuable for the Polish nation. In that same summer Poland's government drew strong criticism from EU officials and human rights activists for illiberal moves to revamp its judiciary after it had already earlier revamped its constitutional tribunal. The European Commission took the unusual step of asking the European

Court of Justice to issue a ban on logging in the Białowieża forest for failing to fulfill its obligations regarding the EU's Habitat and Bird Directives. After the bulk of damage had been done, the highest EU court decided in April 2018 that Poland illegally logged the forest and the door was open for Poland to potentially paying 4.3 million zlote in fines.[2]

But just as humans act upon the forest with their management and designations, there are other beings in the forest who also act. The bark beetle allows us to tell a story about the specter of authoritarian rule in the forest. As the bark beetle defoliates spruce and creates the conditions for PiS to claim the forest as Polish, we see how the Polish national narrative about the forest shifts. The forest's meaning to Poland changes from peripheral in importance to central for nation-building. Discourses converge around authoritarian rule and nationalism vis-à-vis the beetle, where a Polish forest needs Polish foresters extracting timber from the forest. Forests without the Polish forester are not true Polish forests in PiS's rendering.

In eastern Europe, authoritarian rule is most frequently associated with left-wing governments. In Communist Poland (1945–1989), the opposition was imprisoned. There was no free speech and the Communist Party controlled most public discourse. What becomes of this memory throughout the post-socialist period and how does it influence forest politics? As anthropologists we have conducted long-term research in the post-socialist era on the meaning of foresters in Poland. Blavascunas traced the forest conflict in Białowieża, and Konczal studied Polish forester's claims to national interests in Polish forests. We combine our research to reflect on how the myth of the Polish forester has been constructed at the site of Białowieża in PiS's agenda. We are interested in discourses about the Polish past, including both the Communist era and the interwar era, which included many ultra-nationalist sentiments revived today under PiS. It is worth noting early on in this chapter that the Polish Communist party had a flexible and responsive relationship with nationalism in order to support its rule, but it largely suppressed nationalist expression rooted in the prewar Polish narrative of nationalism (Fleming 2010). Thus, our goal as anthropologists in a book with a historical focus is to emphasize how the past is recreated in the present. We also draw critical attention to how non-humans, such as the bark beetle, play into, and even create the possibilities for what kind of stories get told about the forest's past, which in turn matters in what kind of forest is sustained. We do this through a collection of anecdotes and stories that our informants, the people we interviewed and interacted with, have given us. In this case, our informants included foresters, nature conservationists (biologists and intellectuals), nature activists (people actively protesting logging from Polish, local and international circles), inhabitants of the Białowieża Forest region (teachers, farmers, sawmill workers, retirees, business owners), and politicians.

In the immediate decade after the fall of communism, nature conservationists riding on the victory of the anti-communist Solidarity Movement called state foresters "communist-like." For conservationists, the foresters held onto a system of paternalism and favors held over from the Communist era, and thus operated against the then-national narrative of opening to the West, of embracing Western democracy, including accepting Western forms of biodiversity conservation. Yet today, in 2018, foresters and their allies referred to nature conservationists as "communists" as well as "ecoterrorists," all the while emphasizing the esteemed role that Polish state foresters have played in preserving national memory. The post-socialist specificity to this story, in relation to authoritarianism, focuses on how people apply labels of authoritarianism in the dance between nationalism (ultra-right wing, represented by PiS in the present) and communism (the communist era is not the same as who gets called a communist in the post-socialist era).

Employing the ecological metaphor of "periodic outbreaks," we scrutinize familiar patterns in both the bark beetle and authoritarian proclivities.[3] Recently, Polish ultra-nationalists, with a socially conservative platform and an economically interventionist approach, have been using a bark beetle outbreak as a pretext for authoritarian intervention in the forest that extends beyond logging and into policing the left. The left are largely identified with protecting the Białowieża Forest against logging. They call for more EU protections in environmental policy and yet eschew market liberalization *and* reject the former totalitarian state. This is a different left than in the 1990s when former apparatchiks wanted to retain a hold on power in the name of the former system. The left today use the EU's laws to justify their position. By extending an anthropology of post-socialism and an anthropology of post-humanism we engage in what is taking place as we write (2018) and anticipate what is emerging rather than deducing the "facts" of what happened in order to discuss a long-standing bureaucratic figure of the state—the forester.

Post-socialism and post-humanism help us apprehend a world that is unfinished and ongoing, against strict ideas that there are well-defined eras, in which authoritarianism operated or didn't operate. Knowledge and counter-knowledge are being fashioned by political actors and, in the case of Białowieża, especially by nature (ever combined with humans' intentions for what a forest will, should, or could look like). Post-socialist scholarship examines the last thirty years of "transition," mostly in eastern Europe, from state socialism to the retreat of the state in neoliberal forms of capitalism. It has also theorized what appears to be the interminable state of transition in which transition means trauma for those countries that did not succeed in economically stabilizing (Knudsen and Frederiksen 2015; Rakowski 2016). (Poland notably has been called an economic success story, with positive economic growth since 1994.) By comparison, posthumanism is interested in how humans shape historical outcomes. The field questions the hubris of anthropocentrism in a world made up of different assemblages of plants, animals, bacteria, and things, none of which are entirely separate (Smart and Smart 2017).

One overlap between each field is the refusal of determinacy. In other words, post-socialism has refused to believe in a kind of inevitability of capitalism's triumph, or an enduring hold of a communist mentality (*Homo sovieticus*) (Hann 2007; Verdery 1996; Humphrey 2002a, 2002b; Buchowski 2001; Buchowski et al. 2001; Dunn 2004). The social worlds of those in formerly communist Europe changed along familiar patterns. The end of Communism challenged ideas about the nation and the state. Several new nations emerged. Nations that were denied their national histories under Communism, revived stories about the great moments in their nations' histories. A diversity of experience also characterized the way people lived the Communist past and recreated the nation in an era of neoliberal capitalism. Overall, the anthropologists of post-socialism have been able to observe change close-up, but are not "presentists." Temporality matters as does history. In other words, post-socialist anthropology cares about more than what our informants say in the present. It places those up-close interests in a wider historical frame.

By comparison, post-humanism has insisted on the ongoingness of processes between species, between ideas and things, but has less of a temporal and historical element. There is never a rationalist consciousness of (hu) *man*, where one can view the human as an individuated subject capable of sole political authority (Wolfe 2003; Latour 1993). Furthermore, a focus on all the other technologies and beings which make us human challenges all kinds of ethnocentrisms (Haraway 2017; Tsing 2015). No one culture, including culture narrowly defined as humanity, will determine the ultimate outcome of ecological assemblages. In short, post-socialism refuses the determinacy of one type of cultural experience or temporality, and post-humanism refuses the determinacy of the human as the main type of interpretive and agential experience. By using these fields together we demand more than a flat realism of what happened historically, or what a forest is or should be. We do not seek one truth (whether a right- or left-wing authoritarianism can better protect the forest), but rather we seek accounts that are material, political–economic, and ecological.

There is a recurring tension in Poland at the site of Białowieża that cannot go unnoticed in relation to communism and ultra-right nationalism as a historical dialectic. After many decades when the Białowieża forest was widely understood as an "exotic" Belarusian ethnic outpost, the Polish right has claimed Białowieża as theirs. In the winter of 2018, Polish fascist youth groups marched through the streets of the forest's western perimeter, in the logging town of Hajnówka, to exert the supremacy of Poles to this region.

Slavoj Žižek contrasted and broke through this classic notion of competing totalitarianisms (communism and fascism) by highlighting liberalism's role in bringing together right-wing nationalist fundamentalisms *and* radical left emancipatory projects (in which nature conservation activism in Białowieża might be framed). As he writes, "freedom is ultimately *nothing but* the space opened up by the traumatic encounter, the space to be filled in by its contingent/inadequate, symbolization/translation" (2001: 58). For Žižek in his psychoanalytical overview of the post-communist moment, it is the enigma that will break through determinist projects. As we interpret authoritarian proclivities taking shape around the Białowieża forest's management, insights from post-socialism and post-humanism can intervene in both theoretical approaches that search for simplified foils that might echo the solitary narratives of all sorts of authoritarianisms. In other words, our storytelling and analysis in this chapter destabilizes the categories that our informants give us, with the purpose of analysis directed at refracting the tensions that have led to a resurgence of radical nationalism, which we normatively deem unacceptable.

Let us supplement the Białowieża Forest for Zižek's enigma. Both right-wing nationalism and communism are ecological and transformable in our analysis. They are ecological, not in the sense that we as scholars naturalize them, as sociobiologists or evolutionary psychologists might do, but ecological in the way that Isabelle Stegners speaks of when she writes that

> ecology is then, the science of multiplicities, disparate causalities, and unintentional creations of meaning ... Human societies are always susceptible to producing a justification for what they undergo, of transforming their inventions into norms, and forgetting the price paid for their choices.
>
> (Stegners 2010: 34–35)

Putting Stegners into conversation with post-socialism, we turn to an ecological understanding of nationalism in the present moment. Ultra-right nationalisms spread throughout formerly communist eastern Europe because Communism suppressed such nationalisms following WWII, and in Poland the interwar period was already marked by the imprint of Polish culture on Białowieża. The *forgotten price* in this present turn to ultra-nationalism is how interwar and postwar (communist) authoritarianisms denied plurality, and thus led to the replacement of former governing systems, and cultural expressions, by new ones. An ecological interpretation would point to "multiplicities, disparate causalities and unintentional creations of meaning" to make apparent the outbreaks of ultra-right-wing nationalism in relation to the forest, in relation to seizing it as one meaning (Białowieża as a *Polish* forest created by Polish foresters).

Let's look then at ecologies of history, to explain Polish nationalism in relation to communism. In a Polish nationalist narrative constructed by exiles during the Communist period, lived in the Solidarity movement in the 1980s, and idealized since 1989, communism in Poland is widely understood as Russian- (not just Soviet-)imposed. And neighboring Germany imposed its treacherous nationalism over Polish soil many times, but especially in the forest over both World Wars. In that sense there is a kind of psychodrama being played out, a syndrome, doomed to repeat itself. In the

PiS narrative there are no Polish communists and Poles who belonged to the party are traitors, as enacted by the first PiS government's (2005–2007) lustration (outing) of former Communist Party members that held public office or worked in public institutions. The Białowieża Forest became center stage in this drama where the "communists" became "the foreigners" when PiS won the election in 2015. In other words, nature activists defending the forest are called "ecoterrorists" and "communists" as they come to the bark beetle's defense and seek EU support to terminate the logging done to stop the bark beetle outbreak. By bringing attention to spruce bark beetle, and its role in the physical ecology of the forest, we will remind readers of the price paid for nationalisms, not just in relation to the physical composition of the forest, but in relation to the body politic. Polish nationalism has always had to excise Russia, Germany, and communist internationalisms in order to remain Poland, in some kind of pure ideal projection of the nation.[4]

Polish foresters working for state forestry are the central human actors in this story, but we will present them as performative expressions of the state, in its various historical and ecological guises (Scott 1998; Matthews, 2011). Residual characterizations of the forester from multiple eras stay the same in some respects but change in others. Scientific forestry itself catalyzed the idea of nation throughout much of the Enlightenment (Oosthoek and Holzl 2018) and we see in PiS's platform many recycled elements of former expressions of state authority. The spruce bark beetle will enter as an extraordinary agent of history and change in relation to the forester. Clearly, distinct from humans and strategically employed by humans, the beetle has expanded its population at a specific historical moment (post-socialism). The beetle co-creates an undeniable difference in the forest's composition and becomes the pretext for logging intervention that is also an authoritarian intervention against those who do not live up to the ideal of the Polish nation under the PiS government, or who betrayed the Polish nation in the post-socialist period.

To unpack this argument, the chapter is divided into four explanatory sections. The first provides a brief account of the current forest conflict, which escalated in the summer of 2017. This section includes an ecological history of the bark beetle and its role in maintaining a diverse forest ecosystem. The second section locates this conflict in a historical sketch of the forest with special emphasis on the interwar period as it frames the origin story of state foresters and their organization "State Forests" (*Lasy Państwowe*) in Poland and specifically in the Białowieża region. The section also comments on an understudied aspect of State Forests, the communist period (1944–1989) considering this period's hauntedness, or absence in the official history of State Forests as narrated today. The third section provides an ethnographic account of the post-socialist period, drawing largely on Blavascunas's extensive field research in Białowieża (1998–2017). Here we discuss nature conservationists' concern that there were too many "family resemblances" between neighboring dictatorial Belarus (seen as a communist relic) and the Polish communist past. The fourth section returns to the present conflict to identify how and why foresters flipped the national narrative, accusing nature conservationists of being communist and ecoterrorists, and how local youth militias made up of right-wing nationalists led the "creep towards fascism." The article concludes with a post-socialist and post-humanist analysis of how a native bark beetle, when tolerated or even celebrated by nature activists, might deny ultra-nationalist logic and release foresters from the historically narrow role that PiS has given them.

Reversing orders: forest conflict

Naturalists from around Europe long prized Białowieża as a site of relict nature (Daszkiewicz and Samojlik 2014), but Poland's Białowieża Forest took on a symbolic valence as *Europe's* last primeval forest in the years following the political and economic changes of 1989. The area held little status within Poland at that time as a site of Polish national nature. If one looks at the promotional and campaign material for much of the post-socialist period, the Polish Tatra Mountains served the natural national repository for memory. Białowieża had "too many" ethnic minorities to stand as Polish nature. That said, ethnic Polish conservationists seized opportunities, triggered by the changes, to draw attention to the threatened remaining old growth within the forest complex (Franklin 2002). At that time, the nature conservation image of the forest propagated at the international and national level was what one could see in the strict preserve, an ecological assemblage so diverse it is difficult to characterize; a tangle of deciduous and coniferous trees, many of them several hundred years old, patches of swampy forest, and dead wood rotting on the forest floor. From the local point of view, Białowieża had been logged and replanted by them, under the direction of State Forests for nearly a century.

State-supported narratives radically reshaped the way foresters and the Białowieża forest would be perceived. In an abbreviated sketch, Polish foresters formed State Forests as a nationalist cause in the 1920s. In the Communist period (1945–1989) after many foresters had been killed during WWII, for their attachments and loyalties to the Polish nation, the state condemned nationalism. Foresters in the Communist era could not express nationalist sentiments as in the interwar period when foresters were clearly Polish.[5] In the first two decades of the post-socialist period (1989 to roughly 2009), state foresters in Białowieża reactively defended the local rights to resource extraction and development against the urban intellectual elite (Polish interests represented by nature conservation). They also nostalgically defended the former era of full employment and a paternalistic state. The foresters crafted the local into a multicultural set of political actors who could challenge nature conservationists as outsiders, either ethnic urban Poles or internationals. Since the election of the PiS government in 2015 this order has been reversed. The new form of authoritarian-leaning populism reclaims a nationalist and nationalizing past for the Białowieża forest. Ultra-right nationalism is not just about the state controlling resources, but about the performance of one version of the nation emblazoned on the icon that becomes Polish, the Polish forester in Białowieża, who needs to get rid of the bark beetle and foreign interventions.

If the post-socialist period turned an obscure, perhaps forgotten forest into an international icon of a European primeval past, and State Forestry charged itself as guardians of the local, including the Belarusian local, then the PiS government took the symbol of the forest, and through State Forestry, turned it back to the nation. Heavily logging the forest has become a way to challenge the supremacy of the European Union and urban intellectuals, now seen as leftist and even "communist." Local inhabitants who opposed logging once equated conservationists as free-market proponents who wanted to "consume" Białowieża. Now PiS fights for the value of protecting the local as ethnically Polish where logging reaffirms the Polish history that unfolded there. In some ways, logging has become a way to destroy the new left for turning "communist," which happens with the bark beetle at the front of PiS's reason for logging the forest. Forestry "treatments" against bark beetles, including logging and then replanting, are intended to irrevocably transform the ecology of the forest so that nature conservationists can never again claim the commercial forest as their primeval icon. Logging for the PiS government is a short-term prescription for the bark beetle but also serves as a performative act that naturalizes the Polish forester. Logging under PiS accompanies a forest inventory, involving archeology, and forest history that will prove the forest is a human artifact, subject to biblical interpretation of "subduing land" for God's people.[6] The UNESCO World Heritage Program currently lists Białowieża under its natural areas designation, but the PiS government attempts to use these inventories to put the site on UNESCO's *cultural* World Heritage list. In short, foresters under the PiS government are determining the forest in the most modernist of discourses and techniques—through "objective" science (their own experts) and through sanitizing the forest of bark beetle and any ecological relation that would let the beetle's population expand. Dead wood on the forest floor became a "sin" for PiS (Wideo Wyborcza.pl, 2017). But PiS modernist techniques sought to prove their biblical interpretation, connected to their vision of Poland as a Roman Catholic nation.

The foresters' justification for logging heavily at this moment rests on a visible epidemic of infected trees. The spruce bark beetle outbreak of 2017 was not "properly" addressed by foresters during previous outbreaks in the last twenty years, due to conservation measures that mandated dead and dying wood on the forest floor. The beetle found a way of flourishing on the Polish side of the Białowieża Forest with an acute outbreak in 2012 that today's treatments aim to properly address through salvage logging.

Ips typographus is about a quarter of an inch long. It can hibernate for years in downed woody debris. It will emerge from hibernation when conditions are right for flourishing (hot and dry with lots of forest litter). Healthy and weak spruce trees are its food source (Michalski et al. 2004). In the last several hundred years spruce never dominated the forest as much as they do today, which is true for the commercial part of the forest alone. Twentieth-century forestry favored spruce, replanting much of the clear-cut areas from WWI (the first time the forest was heavily logged by modern industrial methods) with spruce seedlings. Climate change, which for Poland means hotter, dryer weather, may extend the bark beetle's range. Infestations are more severe in stands that are over 120 years old, which would be the case in Białowieża, where nature conservationists, many of whom are prominent Polish scientists with the National Academy of Science and several research universities, have long fought for protective forest reserves within the commercial part of the forest. The forest reserves on State Forest lands shelter tree stands over 100 years from logging.

When and how the beetle will flourish exponentially is not entirely predictable, but as with any ecological relationship, the greater the food source and the less the predation, the more an organism will reproduce. Three-toed woodpeckers, a threatened and endangered species in Europe, eat bark beetle larvae, which are deposited under the spruce's bark. The beetles form larval chambers, spreading and eventually killing off their spruce host. A population of three-toed woodpeckers will flourish when the bark beetle population is high, and eventually control its numbers (Bütler and Schlaepfer 2002). The woodpeckers (one of nine *Picus* species that live in the forest, a remarkable aspect of the forest's biodiversity) nest in the cavities of dead trees. Since the mid-1990s foresters and their experts warned of bark beetle outbreaks and set up a trapping and monitoring program. There have been four outbreaks since (1995, 2003, 2008, and 2012–present) evidenced by the volume of dead spruce.

Nature conservation measures secured in 2011 developed a ten-year forestry plan (2012–2021) that mandated that some quantity of dead and dying wood be left on the forest floor. It curbed logging (no more than 48,500 cubic meters a year) and stated clearly that cutting could not occur in "old growth" stands, defined as stands that contained more than 10 percent of trees over 100 years old. In 2016, the PiS government's environmental minister, Jan Szyszko, annexed the Forest Management plan to allow salvage logging treatment for bark beetle prevention and damage. Logging was thus "allowed" in the old growth stands (the legality that the EU challenged).

Biologists and entomologists in Poland and throughout Europe associate the bark beetle as a key species that plays a significant role in natural processes in the ecosystem (Gutowski 2004; Grodzki 2016). A number of prominent biologists agreed that bark beetles could not be eliminated in the Białowieża Forest, nor could their damage be lessened through logging (Kowalczyk et al. 2017). The beetle can fly and when a tree is logged the beetles simply fly to another tree, or they can sometimes be transported to another forest if they are still in their larval stage. The logging in Białowieża took place after the trees were already dead. Unless 80 percent of the trees are removed before they are dead, which could not happen because the National Park does not control for the beetle, the bark beetle will presumably continue to flourish in the future. Some spruce in the forest will die but others can regenerate in the new openings created by the defoliation and toppling of dead spruce (Fase and Heurich 2011; Gutowski and Krzystofiak 2005).

But foresters acted under the requirements of the Polish Forest Act (1991), which in Article 10, Section 1, requires the district forest manager to control for any harmful organisms (Grodzki 2016: 328). National parks are excluded from this article. Foresters argued that bark beetles were attacking healthy spruce because the 2011–2021 forest management plan curbed the harvest. If dead trees are left standing foresters believe they will threaten to all topple over, giving way to a steppe ecosystem, where a thick layer of grass prevents tree regrowth.[7] Thus, the forest will die without the forester and the forest will not re-establish itself if the forester does not replant trees. In the summer and fall of 2017 foresters clear-cut more than 140,000 trees, 35 percent of them over 100 years old and removed dead and dying timber from the forest reserves. In the spring of 2018, foresters replanted trees while biologists argued for natural regeneration. Unlike the Białowieża National Park, which has a scientific council made up of many different kinds of experts in society, such as biologists, foresters, historians, cultural experts, and professors, State Forestry only uses their own experts and they are accountable only to themselves, not to any kind of overseeing body of experts. Poland has a strong tradition of university-based forestry programs, which are not all in agreement with the PiS government's approach to the bark beetle within Białowieża, but few foresters in Poland dare speak out against the current administration for fear of losing their positions (Robiński 2018; Książek 2018). PiS roundly purged state-owned media in 2016 and some foresters fear the same for themselves.[8]

Białowieża has had a small local group of activists in the post-socialist period who fought for the forest's protection against logging in the 1990s and early 2000s. The activists were supported by larger national and international groups, such as Greenpeace and the World Wide Fund for Nature (WWF). However, since the changes to the forestry plan were announced, hundreds of activists, many with no affiliation with any group, have encamped in the forest villages. They regularly monitor the logging, using hand-held GPS units to mark every tree slated for logging, even after State Forestry forbade the public from entering the forest in the spring of 2017. They have repeatedly blocked logging equipment by chaining themselves to harvesting equipment. In response, the chief of State Forestry, Konrad Tomaszewski, called in a special paramilitary unit of foresters with the right to carry guns in the forest, called "Forest Guards." Foresters began using strong-arm and militaristic techniques in the summer of 2017. They wore protective bulletproof vests, despite the peaceful nature of protests and blockades. They confiscated the media's cameras and footage. They dragged protesters from equipment, applied body-locks, arrested protesters, and fined them heavily.

For protesters and other nature conservationists, the bark beetle is a symbol of a new kind of ecological success story. It is their story of rewilding the forest and finding new ways of socially relating in a civic democratic fashion. Protesters pride themselves on being part of an international movement that also eschews nationalism. They base their camp on anarchist organizing principles familiar to the Occupy movement that started at Wall Street and spread around the world in 2011. There are no leaders and no followers, only spokespersons (like the spokes of a bicycle wheel) elected by councils of those present. There are no logos (such as Greenpeace or WWF). They see themselves as a people's movement. Decisions about when and how to set up blockades are set by the collective, which grows and shrinks depending on different external factors, such as people's work schedules, family commitments, and monetary resources.[9] PiS and its supporters saw the protesters as a swarm of unruly people. They have called the protester's "pests/*szkodniki*." Sometimes the protesters do not speak Polish, making commands and communications difficult as forest guards bark directions to dismantle their blockades.

If the protesters deny nationalist logic then the foresters, under the direction of PiS enforce nationalist logic. In the nationalist logic, minorities and people who do not fit the national ideal always reside within the national territory and need to be kept in check. In times of economic or other stress, the nation needs to excise the "foreign" elements to create solidarity. Nationalists dig into their glorious pasts (victories of war, of territorial gain) and their victimhood (defeats in war, impositions of other or imperialistic powers) toward what they feel is the one and only solution to injustice—the nation. What is the ethnic history of the forest though that created such a strong reactionary nationalism in 2017, a reactionary nationalism that called nature activists "ecoterrorists"?

Ethnic history of the forest

When Poland regained its independence in 1918 (after a partition between Russia, Germany and Austro-Hungary in the eighteenth century), the Białowieża Forest had been part of Russia for more than 120 years. Prior to that it belonged to Poland, being first protected as a royal hunting ground in the fifteenth century when Poland and Lithuania united. Under nineteenth-century Russian rule, Polish magnates lost their land. Local residents, who had been local forest guards and beaters under Polish rule, were forcibly converted to the Russian Orthodox Church, and Polish Catholic Churches dismantled (Daszkiewicz et al. 2012). Many Polish gentry in the nineteenth century wanted to apply the fashionable new science of forestry to their estates. Polish forestry in the nineteenth century was a bastion of discontent against Russian rule, so much so that Russia shut down the forestry school in Warsaw after the 1831 uprising (Majchorwska 2018: 325–331). In short, forestry science in Poland would have been a place for organized Polish rebellion in partitioned Poland. The landed Polish aristocrats committed themselves to forestry to place themselves on a tract of national modernization.[10]

During WWI, German occupying forces began logging the forest, using industrial equipment and methods for the first time in the forest's history. They built an industrial complex for logging at the western rim of the forest in what was once the village of Hajnówka. From 1921 to 1928 the British Century European Timber Company (hereafter Century) logged on contract for the Polish state. Century used and owned the wood, sending the milled lumber back to the UK. Many ethnic Poles were brought in from central Poland to work in the mills as the local Belarusian-speaking population became identified as susceptible to Bolshevik propaganda and were not hired for work (Nikitiuk 1958; Bajko 2017).

The break with the British contractor came as Polish foresters sought to terminate what they saw as an exploitative foreign relationship. State Forestry's first director, Adam Loret broke a contract that was costly for a new nation needing the timber revenue, but as foresters today narrate this story, Loret acted to protect *Polish* forests. Loret would later be killed by the Soviets in 1939. Loret's example shows that State Forestry was founded to take the forest into their care against foreign powers, regardless of the monetary cost (Broda 2006a; Grzywacz 2014). Loret's death is today celebrated as a martyrdom with a new statue erected to Loret in 2017 outside State Forestry headquarters in Warsaw, as well as a monument in front of the Białowieża Technical School of Forestry, the *Odsłonięcie pomnika* built in 2017. The statue speaks to the PiS platform to elevate socially just causes over economics.

The Białowieża National Park was created by one of the founders of the International Union for Conservation of Nature, a Polish biologist from Lviv, Wladyslaw Szafer (Okołów 2009). Szafer surveyed the damage to the Białowieża forest (first by Germans and then by Century) and set up the core nature reserve that would become the Białowieża National Park in 1932. One of the first protective reserves on the commercial forest was named "The Szafer Reserve."

But it is the foresters and not the nature conservationists who today become lionized as martyrs for their roles in WWII. Germans occupied the forest in 1939, but this time, under the direction of Reichsmarshall, Herman Goering, who preserved the whole forest for its Teutonic qualities. He envisioned a large nature preserve with no logging, where hunting would be allowed. Following the German occupation, Soviets occupied the forest in 1941and logged the forest for its timber (Okołów 2009). At that time Belarusian inhabitants identified ethnic Polish foresters to the Soviets who then carted the Polish foresters to Siberia. Many Polish foresters fought as anti-Soviet partisans during the war. The war took its toll on Polish foresters who died in great numbers. Because they were educated, Polish foresters were often officers in the war. The NKVD (People's Commissariat of Internal Affairs) secretly murdered more than 22,000 Polish officers during the war at a mass burial site in Katyń Russian, many of them Polish foresters, a historical betrayal popularly remembered today as the "Katyń crime" (Peszkowski and Zdrojewski 2001).

After the war, when Churchill, Roosevelt, and Stalin split the forest in two under the Yalta agreements, Communist Poland changed the story about the struggle for the Polish nation within the forest. The Party, consistent with Communist ideology, denied national groups their histories. Nationalism served bourgeois interests. Thus, there was no distinction made between Poles and Belarusians regarding work in the forestry sector (though ethnic Poles continued to dominate the higher administrative positions within State Forestry). Throughout Poland and communist eastern Europe, former estates (many of them forested) were broken up. In 1944, as the Red Army moved west and the Soviet Union claimed an independent communist Poland, land was redistributed and in many cases nationalized. Forests went to the state (Broda 2006b). As many Belarusians had been peasants in the forest region, and farming land was rimmed in by the great forest, State Forests advanced poor peasants, irrespective of their ethnic identity and Poles and Belarusians intermarried, blurring ethnic lines during the Communist period.

Post-socialist anti-authoritarian forest politics

In the early 1990s, Poland and Polish State Forests downsized their labor force. On the basis of the 1991 Forest Legislative Act, Polish forestry adopted a corporate model, meaning that the organization was self-financed. It kept almost 80 percent of state-owned forest under its supervision, meaning that one-third of the territory in post-socialist Poland was directly managed by State Forestry. It also ensured an almost complete monopoly on timber sales.[11] Only since 2014 has State Forestry been required to pay 2 percent of its income from wood sales into the national treasury. Despite this corporatization of State Forestry, many nature conservationists during my (Blavascunas's) ethnographic research in the late 1990s made the point that State Forestry was the "last authoritarian organization in Poland." It had a "communist-like mentality" and way of organizing that thwarted all attempts to expand the national park or better protect the forest as a whole.

In this way, the post-socialist conservation movement that began in the early 1990s formed around the suspicion that foresters were acting in communist/authoritarian ways. Conservationists remarked on the following: foresters wore uniforms on and off duty. They controlled the sale of heating wood and thus the local population's support. They hunted with politicians to get their agendas passed. They ridiculed their opponents. In short, none of these behaviors in and of themselves meant foresters were "communist," but at that time the perception was that communism discursively meant a lack of transparency and conducting business in the "old way," rather than changing the rules for social conduct. The resemblance between the Białowieża forester's power and neighboring Belarus, widely understood as a communist country, was too strong to go unnoticed. Nature activists made many comparisons between Alexander Lukashenko, the former collective pig farm manager who became president of Belarus in 1994 (and remains in power today), and State Foresters and the way the region operated. Belarus suppressed freedom of speech, jailed opposition candidates, continued to use Soviet era memorabilia and retained streets named after Marx and Lenin. Belarus designated their entire forest as a national park in 1991. However, the forest though had received better protection from logging prior to the designation.[12] Foresters ran the National Park and logged the forest, including sanitizing the forest against bark beetles. Here, I (Blavascunas) recorded how nature conservationists and a visiting Belarusian nature activist, framed foresters in relation to Belarus, as a relic from the Communist past) (parentheticals by Blavascunas):

VASSILII: (Belarusian nature conservationist): I have not been allowed to go to the *Puszcza* (Russian and Polish word referring to Białowieża Forest) for two years. Lukashenko owns the forest and told me it is forbidden to go (because of Vasilii's activist work against current management of the Belarusian forest). But the forest is big and there are other ways to access it. (He grins with a sly satisfaction). After this ban, I organized a press journey of both the legal and underground press. We were all organized and planning to leave. At 6:00 a.m. they (KGB) called me and told me: "You have seen enough of that forest!" They didn't allow me to go and took the press on their own tour of the forest. You are lucky to have a better situation on the Polish side.

JANUSZ: (Polish nature conservationist): I'm not so optimistic.

VASSILII: If we change the democratic situation in Belarus we can change the situation in the *Puszcza*. Most of our problems will vanish with democracy.

ADAM: (Polish nature conservationist): Don't you understand! The Białowieża Forest region is little Belarus in a pill. The foresters here can do whatever they want. Białowieża Forest received the European Diploma (a designation from the European Council of Protected Areas). Instead of this they should have given Poland a sawmill diploma.

On the Polish side of the forest, the post-communist period saw a short period in which Belarusian identity flourished. Yet what was Belarusian identity within Poland, given present-day Belarus (where Russian and not Belarusian was the official language)? In the post-socialist era, Belarusian identity was promoted as somehow more native, a point much exploited by foresters (as well as nature conservationists) to retain status and appear as truly local, as multicultural and open in the newness of the 1990s. In the first major protest in 1999 against national park expansion, foresters organized locals to hold signs in Polish and Belarusian calling conservationists "Judas traitors" for their wishes to use the EU to protect the forest from logging. The EU was somehow seen as selling out Belarusian interests. Belarusian identity activists (within Poland) wrote articles claiming that a huge national park would destroy the Belarusians' chances of modernizing (Fleming 2010; Blavascunas 2010). Belarusian identity activists claimed that Belarusians in Poland were poor and needed logging to modernize themselves and the region (the development of underdevelopment). Many locals, including Belarusians, Orthodox Christians, Poles, and Roman Catholics supported foresters. There was no objection to being compared with Lukashenko and Belarus, because he was not a ruthless dictator in their eyes, but an efficient manager, who might use the strong-arm tactics of the Communist days to achieve a greater good for Belarusian society. "We are a society that needs to be ruled by an iron fist," one mixed Polish-Belarusian inhabitant of the Polish side of the forest told me. Foresters and locals brought up comparisons to Belarus as a better functioning place. In Belarus, the Forester had more respect, as evidenced by his ability to sanitize the forest from bark beetles. A local Polish pro-forestry lobby circulated media of Belarusian foresters coming to Poland and announcing with disgrace how irresponsible it was of Polish conservationists to legislate that dead wood be left to rot on the forest floor.

In the first two decades after communism, it did not matter that Lukashenko was in power for the local people in Białowieża. Logging and a Belarusian or Polish identity could be one and the same. It did not matter for the logging lobby or the forester, if people were Polish or Belarusian. All that mattered was the continuation of logging with the forester in charge. This was the transhistorical power of foresters. They could be ethnic purists in the interwar period, ethnic unifiers in the communist period, or ambivalent about Lukashenko in the 1990s and 2000. But for a century they held power in the region and decided the fate of the forest. PiS acted in this "all powerful way" now. The authoritarian government then did not need to be embodied in a strong "man" but could appear in the strong arm of the state as embodied in foresters who acted expediently, on popular mandate, to crush their opponents.

But for the conservationists and nature activists, logging culture associated with Belarus represented a time lag. The Polish foresters' and Belarusians' approach to bark beetles was outdated. Ecological studies coming out of places like the Pacific Northwest in the United States validated the role of dead and dying timber in the ecological cycle of the forest (Harmon and Hua 1991). Long-term research at the Sumava/Bavarian forests also proved that a coniferous forest could regenerate better if logging was not prescribed for a bark beetle attack (Jonasova and Prach 2008; Petrova 2016). According to biologists, bark beetles were *European* and could bring back an "original" nature without causing forest die off. Belarus was like the old Communist Poland, the old Soviet Union. Thus, the locals on the Polish side of the forest had a "communist-loving" mentality, according to biologists. The way in which foresters spoke of their paternalistic mission to care for all local people, to revive the timber economy and provide everyone with jobs like in the good old days of communism, suggested a possibility of a future socialist state. Foresters acted against nature conservationists who scaled-up to get support at the level of Europe and the international community. But the ecology of this narrative, of forester discourse, and of the forest changed with the rise of Polish nationalism following the PiS electoral victory in 2015.

Targets of the ultra-right

Since PiS was founded in 2001 by Jarosław Kaczyński (twin brother of former President Lech Kaczyński), the party has been dominated by the Catholic Church and social conservatism. Starting as an opposition party they won their first election in 2005, then lost to Civic Platform in the 2007 elections and won again in 2015. In 2017 the bark beetle and its cyclical outbreak became instrumental in PiS's platform. The beetles fit within PiS's narration about "threats" to the Polish nation, which have become threats to Polish nature and Polish forests. Scheming foreign powers are always waiting to attack Poland, proverbially and literally for PiS. Thus, the bark beetle makes it easier to identify the current enemy of the Polish nation. This time, the old foes (Communists, Bolsheviks, Soviets, Russians, German, or British logging interest) appear in the new face of conservationists and ecological activists. Green activists become traitors of the Polish nation, who serve foreign affairs and peddle foreign values. This nationalist obsession reaches far beyond the Białowieża forest. On the wave of the right-wing sentiment, all who cannot fit into the pattern created by PiS including ethnic minorities, religious minorities, sexual minorities, and feminists, have to be detected and excised, a political configuration PiS refers to disparagingly as *lewatcwo* ("lefties" or "pinkos). The outbreak of the bark beetle is then an excuse for capturing not only a definition of the correct nature management but also for establishing the one and only proper history of (and for) the nation.

Before being appointed as the Minister of Environment in 2015, Professor of Forestry, Jan Szyszko said:

> We are preparing the forest law, which will protect forested land. This land will be protected by Polish law. By that, this land will be serving Polish people; will be a guarantee of the existence of the Polish state and the Polish nation. (…) The nation without land is a dead nation.
>
> (Szyszko, 2015)

The forest, and especially the Białowieża Forest, began to be perceived and presented as a proxy of the Polish nation. In this narration, no foreign interference should influence Polish forests, especially on the matter of the bark beetle. Nature conservationists, the international law and order exemplified by the European Union or UNESCO in its condemnation of Polish logging in Białowieża were all foreign.[13]

In Szyszko's statement, Polish forests must be under the control of only Polish foresters who are servants of the state and the nation. Thus, foresters need a particular version of history. The Polish nation *is* the story of past forests in Poland and past foresters. This forest myth is based on an assertion that there is no Poland without forests and no forests without the Polish-ness of its foresters. Foresters set themselves up as the heroes of this myth. In the most difficult historical moments, such as the 1831 and 1864 uprisings, WWI and WWII, Russian or German occupation, foresters led and saved the national soul by saving its forests. The patriotism of previous generations of Polish foresters, especially those from the interwar period, is presented and interpreted by contemporary foresters, as sacrifices leading to the current political outcome—Poland as an independent country with safe and healthy forests.[14] A past forester's activities, in logging and planting, administering and rebelling, explain today's legacy, and at the same time they are tested schemes of practice, which can be used again.

The myth needs to be performed for its historical continuity, which happens in two ways. First, history from the forest is not based on that which *is* within the forest (for that would need to include old growth and bark beetles); it is built upon that which is precisely *excavated* from the forest (through forester's management and labor). Second, both the scientific findings (through archeological excavations) and Polish forestry science on bark beetles, need to be performed by fighting "ecoterrorists" who block their harvesting equipment and seek extra-Polish legal means of nature protection. This performance of forest history takes place in response to, and in consort with, the actual version of the national history. Notwithstanding, the character of Białowieża provides tools and materials for this performance (the bark beetle, the repeated foreign occupations, the martyrdoms vis-à-vis graves in the forest, its Belarusian otherness). Foresters do not build the "national history in the forest" according to their own will, but borrowing from post-humanist scholar Anna Tsing, on the basis of a "set of reminders from the past that we turn into stories" (Tsing 2015: 168). Foresters are adept at calibrating local long-term opposition to nature conservationists to meet their national forest myth. With PiS, Białowieża's stories have again started to serve nationalistic purposes of the state. With the outbreak of the bark beetle the forest myth became a nationalistic story.

By about 2012, Catholic media, ultra-right wing and nationalistic in its programming, took up the cause of Białowieża. Both Radio Maryja and TV Trwam broadcast Polish nationalist religious messages that are at once rosary-specific as well as anti-feminist, against the Islamization of Europe,

Middle Eastern refugees, and LGBTQ rights. The radio station heavily used the term "ecoterrorists" to refer to groups that actively opposed logging Białowieża. Jan Szyszko, both before he was appointed Minister of Environment, and after, appeared many times as a guest on Radio Maryja and TV Trwam, where he often spoke about the need to assert the proper (forester) control over Białowieża against those who would let their ecoterroristic vision of strict protection destroy the Polish forest.

By 2013, a Hajnówka-based group calling itself Santa, funded by State Forests and whose President, Walenty Wasiliuk (an ethnic Belarusian) owns several sawmills, organized a protest in which the first real signs of Polish nationalism appeared as the dominant driver of logging on the pretext of bark beetles. Their banners read "Save the *Polish* Forest from Pseudo-ecologists and those who would sabotage it." There were people dressed in folk costumes from other parts of Poland, and Catholic as well as Orthodox priests appearing together with Szyszko on stage, including the regional chaplain of foresters, Tomasz Duszkiewicz, who stated that anyone who criticized hunting was "not Polish." Radio Maryja and TV Trwam constantly streamed messages about the necessity to log the forest in connection with stamping an ethnic, Catholic Polish identity.[15] In this performance, Belarusian appeared as a regional identity, subsumable to the larger Polish project of claiming the forest for Poland.

Although there have been nationalist elements of many post-socialist governments, the PiS 2015 victory enabled expressions previously unseen in Poland, such as banners for a "Muslim Holocaust" unveiled in a 60,000 person-strong march on Polish Independence Day[16] and white supremacy demonstrations protected by Polish police. However, in the whole Białowieża region the party failed to win majority votes and only in the singular Białowieża commune did PiS win the election. Białowieża's local attitudes to the ruling party have changed quickly together with a visible rise in the nationalistic approaches among local inhabitants. When Jan Szyszko was nominated to Ministry of Environment he made it his personal mission to radically transform the ecology of the forest, through an attack on bark beetles, and the message about the importance of Białowieża to the national identity. A few months later Szyszko dismantled the National Council on Nature Protection, most of whom favored a national park expansion or backed the studies that the native bark beetle could architecturally reconstruct a forest misdirected by forestry logic. Szyszko dismissed thirty-two out of thirty-nine members and constituted a new Council of twenty-nine members, among them mostly foresters, agricultural scientists, and directors of botanical gardens.

In the winter of 2016 European Union bodies watched Poland, criticizing its moves to dismantle its constitutional court. Later, in July 2017, the EU gave Poland the ultimatum to drop judicial reforms that would have allowed PiS to put judges under its political control. If Poland did not comply it might lose its voting rights in the EU, the ultimate "nuclear" option. In this context, the clash between competing narrations about Białowieża became a fight for the nation and its ultimate values. PiS is trying to confirm its version of the nation, country, nature, and the forest.

Belarusians (and other national, ethnic, religious, and sexual minorities) increasingly became the target of local ultra-nationalists, who took to celebrating the "cursed soldiers" (*Żołnierze Wyklęci*). "Cursed soldiers," a term invented in the present era, refers to those who fought in Poland's "Home Army" (*Armia Krajowa*)—the dominant Polish resistance movement under WWII occupation. The "cursed soldiers" refused to put down their arms after "liberation" by the Red Army. They chose to fight as partisans against "illegal occupants," meaning the new Soviet-imposed rulers, and constituted anti-communist Polish resistance movements between 1944 and 1946 (but some units operated until the early 1950s). However, many of these soldiers died during fights with Soviets and Polish Communists or were sentenced to death under Communist controlled courts. Some of these soldiers were responsible for revenge killings against Belarusian inhabitants who identified Polish foresters to Soviets during the Soviet occupation of eastern Poland. Communists banned any mention of these units.

Post-socialist ideological discourse revived the national memory of the cursed soldiers, with monthly journals dedicated to the fallen. Such narrations have only increased with PiS rule, which legally mandated the inclusion of the cursed soldiers in the elementary school curriculum.[17]

The creep of fascism, which may or may not lead to a more conspicuous totalitarianism in Poland, is that non-governmental organs adopt the extreme and literal interpretation of the government's nationalist position. The popular election of PiS speaks to the uncoerced mobilization of a part of the population. But since PiS's recent rise, the most influential organization strongly reviving the memory of "cursed soldiers" is the National Radical Camp (*Obóz Narodowo-Radykalny*, ONR), a far-right non-political movement which considers itself a descendant of the political movement from interwar Poland (which shared the same name). The first ONR was established in 1934 by young radicals (anti-communist and nationalists) and it was influenced by the ideas of Italian fascism. The ONR called for the elimination of national minorities from public life. In 2012 it was legally registered as a common-interest association. The National Radical Camp uses an international racist symbol, the Celtic cross, to advocate "white power." They promote anti-Semitism, xenophobia, homophobia, and use strong anti-refugee, anti-Islam language. They are accused of using violence against minorities, as well as anti-governmental activists.[18] On April 29 and August 15, 2017, the ONR organized marches in Warsaw where they carried fascist symbols and raised their arms in "Heil Hitler" salutes. At all of these events, police escorted members of the ONR and removed anti-fascist protesters. Although marginal, the ONR attracts many who have worked in western Europe and oppose a secular Poland. It is not uncommon for both the PiS nationalist agenda and the ONR's to appear at "pluralistic" events dedicated to commemorating the Polish nation.

In 2017, local activists of the ONR associated also with the "Narodowa Hajnówka" unit (*National Hajnówka*) entered the Białowieża Forest conflict. They support Szyszko and State Forests, calling for more intense and expedited logging—in the name of bark beetle extermination. Members of this organization attended blockades in the forest and verbally threatened conservation activists, perceived by ONR as enemies of the nation. Their Facebook page, iconized as a slashed through hammer and sickle, called upon members to form militias that would track down "terrorist" activities, referring explicitly to nature activists. Young Narodowa Hajnówka activists (who sometimes come from mixed-ethnic and mixed-religious families) performed PiS's narration, which connects the anti-Russian story of "cursed soldiers" with "ecoterrorists" and bark beetles. Their fight with enemies of the nation is based on the one correct version of the history and forest management. Wearing stickers with the slogan "Death for Enemies of the Fatherland," they argue not only for cleaning the forest from bark beetles, but also for cleaning the nation from "foreign factors," such as the encamped activists.

In a current political context in Poland local foresters together with fascist youth want to tell the story of Białowieża as a *Polish* forest, which is a symbol of Polish land, Polish heritage, and Polish national values. In this story foresters present themselves as former partisans, fighting against communists, in the name of "great Poland." They present themselves in victimhood against Russia and the Soviets: the Russian tsar sent them to Siberia, and then the Soviets killed them during WWII and during the anticommunist uprising. But in the current context, the enemy is linked with any kind of otherness—it can be associated with Russia, communism, terrorism, Islam, feminism, ecology, or anti-governmental movement, and therein lies the danger. Their promotion of one true narrative denies the plural narrative of a multicultural forest with a complex ecological relationship. Historically, this limited narrative in the name of nation led to violent and murderous outcomes in the twentieth century.

In June of 2017 foresters erected a statue to one of the cursed soldiers within the Białowieża Forest, a ceremony attended by those from Narodowa Hajnówka, who chanted "*A na drzewach zamiast liści wiesić komuniści*" ("Instead of leaves hanging from trees, there will be communists"), a popular cry of the radical right at nationalist marches. As the statue's likeness, then eighteen-year-old Danuta Siedzikówna, who worked for State Forestry in Hajnówka, has become an icon of the "cursed soldiers" movement. Her father was captured and sent to Siberia by the Soviets in 1940 and her mother was executed in the forest by the Gestapo for collaborating with the Polish Home Army, the dominant Polish resistance group in German occupied Poland. She and the statue symbolize a renewed fight with anything that looks outside the Polish narrative of victimhood, Polish narration about history, its heritage, and value. This fight includes the bark beetle—the pest inside the Polish forest and the enemy of Polish foresters—as well as the fight about whose ecology and science count.

Simultaneously in the PiS and ONR narrative, ecoterrorists are the bad part of the EU—the worst version of a decadent secular West. Ecological activists have used arguments about the forest as "the heritage of humanity." They have garnered the support of the international conservation community and have appealed to the European Union, its Court of Justices and other international agendas—instead of the Polish government and the Ministry—as institutions having juridical and legal power over Poland. They were scaling up, while Polish foresters, immersed in a new nationalistic narration, were scaled to the Polish nation. What appears to be happening now in relation to the emergence of the ONR in Białowieża, and the ethno-celebration of Polish foresters is that the ethno-Polish forester, in militarized anti-riot gear, will take care of the swarm. The ethno-Polish forester will take care of the swarm that is the bark beetle, but also the swarm that has become the nature activist, seen in the on-the-ground battle for the forest as non-native, foreign, non-local, and non-Polish—as the Other: the enemy of the forest as much as the bark beetle.

Ongoing processes

By recognizing the beetle as part of the forest's ecology, attention is placed on processes, rather than a stagnant view of a beautiful forest with old trees, or narrowly on charismatic species, like wolves or bison. The bark beetle is an ally then, a small unsightly insect that can do drastic ecological work. It can diversify monocultural forest stands and topple ancient spruce in a national park. When one cares for the bark beetle and considers its role, one cares for the forest as a series of ongoing, dynamic processes. These processes make people care about the forest for its ecology and its relationships with life and death, eating and being eaten. The forest has been evolving for 12,000 years, since the end of the last ice age. In 99 percent of its evolution the forest was never cleared for agriculture. Attentiveness and allowing the presence of bark beetles then rearranges the icons of the forest. In this framing, Poles are taken away from narrow nationalist usages of forest symbols, such as the bison (an animal long symbolizing the ruler's right to hunt and rule), when they attend to the bark beetle's creative evolutionary role.

If we apply insights from post-socialism and post-humanism together, to look at the ambiguity of the forest and the way authoritarian-leaning governments seize the forest's meaning, we see a temporal configuration. There is a way of organizing time on the part of authoritarian-leaning regimes, which contrasts with ecological time, even as the authoritarian-leaning regimes seek to imprint their temporality on nature. Evolution unfolds as a slow process inconceivable to the naked eye, and only knowable through an exploration of deep geologic time. An authoritarian state, such as any Communist state, can seize people's relationship with time to make them believe in one dogmatic narrative. The two fields, post-socialism and post-humanism, both divert attention from mainstream temporal narratives that want to see radical breaks between historical periods, or between nature's time and human's history, between human knowledge and non-human passivity (was socialism/now capitalism, individuated rational knowing subject/collective action through a network or assemblage in ecological relation or metabolism).

The turn to nationalism in Poland is a shrewd discerning of the ruptures of internationalism, of fast capitalism, of the failures of "transition." Politicians have diagnosed a discontent of belonging to the European Union in favor of belonging first and foremost to Poland. If a European-looking government could not care for its people, then the inward nationalist government of PiS would. The Polish government must care for its trees, must mark the otherness of a native beetle to win the sameness of the Polish people.

There are processes by which places, ideas, cultural groups appear threatening to the grand design of nationalist projects. These are places and phenomena and cultures that do not fit the design to perfection—they are ambiguous. Yet, in part, PiS itself relies on a kind of ambiguous distance between its official position and its tolerance and even complicity with ONR. An analogy between nativisms (bark beetles and ethnic or political otherness) allows us to challenge what is considered threatening and what is considered significant. By dint of comparison we have a critical vantage point to challenge claims about hierarchy of value.

In a moment often called the *nationalist return* in Europe (yet a phenomenon clearly operating at a global scale), a turn that securitizes belonging through protection of frontiers, an analysis of the post-socialist period using a combination of post-humanism and post-socialist scholarship teaches us about the ongoingness of processes. It also provides a temporal optic, which lets us think about historical change and possibilities. In using the narrative and agency of bark beetles we radically rethink relations of production, both historical production, national production and forest production; how the nation is produced through natural resources by authoritarian-leaning governments. Once we see the bark beetle as part of a plurality, of radical difference, the way post-humanism asks, and trace a history of that difference in social relations too (one of the concerns of post-socialism), we are opening ourselves up to a way of becoming that might learn from all the ways the twentieth century has tried to stamp out difference, or narrow identities and forest diversity.

Notes

1 The Second Republic of Poland (1918–1939), PRL (1945–1989), Third Republic of Poland (1989).
2 www.theguardian.com/world/2018/apr/17/poland-violated-eu-laws-by-logging-in-biaowieza-forest-says-ecj.
3 Outbreaks of nationalism are certainly not limited to the interwar period and today. One could point to pogroms during WWII or the expulsion of Jews from Communist Poland in 1968, but our framing wants to look at the specificity of Białowieża in the interwar period and today, especially related to the ethnic Belarusian minority in the region.
4 A long-standing Polish national narrative has been built on the idea that Poland is the "Christ of Nations" or that Poland is the perpetual and misunderstood martyr and victim in Europe, where Europe never fully understands Polish contributions to European identity (Davies 2001).
5 See Francine Hirsch for a discussion of nationalisms within the Soviet History's development (2005).
6 In 2016, State Forest's Institute for Forest Research inventoried the cultural heritage of the Białowieża Forest. Through archeological data the Institute documented ten centuries of human use. www.ibles.pl/en/web/guest/badania/puszcza-bialowieska-dziedzictwo-kulturowe/inwentaryzacja-dziedzictwa-kulturowego.
7 www.bialowieza.bialystok.lasy.gov.pl/edukacja/-/asset_publisher/1M8a/content/spruce-bark-beetle-in-the-bialowieza-primeval-forest.
8 In 2016, hundreds of journalists were fired or quit their posts at State broadcasting outlets. www.nytimes.com/2016/07/04/world/europe/polands-conservative-government-puts-curbs-on-state-tv-news.html?_r=0.
9 "We call for mass grassroots mobilization. Let's show that the Forest belongs to us all, and we can defend it together," "We organize a camp opened for everyone, who cares about the fate of the Forest," "[We are] Building a grassroots coalition for the Forest. From foreign activists, to concerned locals—we support each other and cooperate to protect this unique Forest," Forest activist website: https://pomagam.pl/dlapuszczy.
10 Forestry came to Poland through Austria and Prussia. Through forestry schools, forestry provided the infrastructure that was used by Poles as organizing nodes where they could prepare their nation-liberating activities. This situation refers also to the general movement in partitioned Poland, concentrated on "healing the country by the work and sacrifice." It was strongly impacted by the period in Polish literature called "pozytywizm" (positivism) with novels by the most important national writers like Henryk Sienkiewicz, Bolesław Prus, Stefan Żeromski.
11 The forest ownership structure ensured by the Act about forests from 1991, provides an almost complete monopoly on timber sales to the State Forestry. In turn, this allows foresters to dictate the terms to the market in regards to the price and the amount of wood sold. State Forest controls 80 percent of the domestic wood market.
12 Soviets initially declared the area a Zapoviedniki in 1939, but then in 1957 Soviet officials removed that status and it became a State Protected Hunting ground. Although they drained much of the forest to create two fishing lakes, they refrained from logging.
13 In a 2016 statement to the Polish Sejm, Szyszko said

> Forest land is in this situation unprotected. And now, I want to tell you all, I want to tell you that at some point, we come back to the era of the Second Polish Republic. Then, just after getting the independence, the base for the

> State Forestry was created (…). The poor Polish State bought 800,000 hectares of forest from private hands. They did this to prevent logging in these forests. To prevent these forests from getting into foreign hands! In this moment, with this act, we propose the same. Thus, ladies and gentlemen, we are in the similar situation as in the time of the Second Republic of Poland.
>
> (Transcript from Parliament Meeting No. 16, 2016: 197–198. Parliament Cadence No. VIII (2016). Prace Sejmu VIII kadencji)

14 Poland has one of the most carbon intensive economies in Europe, burning (Polish) coal for 90 percent of its energy. The environmental minister, Szyszko, has pushed for Poland to use its "healthy forests" as a carbon offset in climate talks.

15 https://oko.press/zielone-ludziki-ministra-szyszki/.

16 www.independent.co.uk/news/world/europe/poland-far-right-demonstrations-great-celebration-warsaw-neo-nazis-white-supremacists-a8053516.html.

17 On April 10, 2010, an aircraft of the Polish Air Force crashed near the city of Smolensk (Russia) killing ninety-six people. Among the victims was the President of Poland Lech Kaczyński, brother of Jarosław (current PiS chairman). The people on board were arriving to attend an event marking the 70th anniversary of the Katyń crime. PiS claims the crash was a political assassination by Russia.

18 In Radom, the city south from Warsaw, during the anti-governmental demonstration, members of the Committee for the Democracy Protection (NGO) were physically attacked by activists from ONR and other nationalistic units *Młodzież Wszechpolska* (All-Polish Youth) who were openly supporting PiS.

References

Bajko, P. 2017. "Rozstanie z Olga Gabiec." *Czasopis*. http://czasopis.pl/rozstanie-z-olga-gabiec/.

Broda, J. 2006a. "Główni architekci organizacji i gospodarki lasów państwowych," in A. Szujecki (ed.), *Z Dziejów Lasów Państwowych i Leśnictwa Polskiego 1924–2004, Vol. I: Okres międzywojenny*. Warsaw Centrum Informacyjne Lasów Państwowych, 113–139.

Broda, J. 2006b. "Leśne dziedzictwo po okresie zaborów i powojenne uwarunkowania," in A. Szujecki (ed.), *Z Dziejów Lasów Państwowych i Leśnictwa Polskiego 1924–2004*. Warsaw: Centrum Informacyjne Lasów Państowych, 29–97.

Buchowski, M. 2001. *Rethinking Transformation. An Anthropological Perspective on Postsocialism*. Poznań: Wydawnictwo Humaniora.

Buchowski, M., Conte, E., and Nagengast, C. 2001. *Poland beyond Communism: Transition in Critical Perspective*. Memphis, TN: University of Memphis.

Bütler, R. and Schlaepfer, R. 2002. "Three-toed Woodpeckers as an alternative to bark beetle control by traps?" International Woodpecker Symposium. *Nationalpark Berchtesgaden Forschungsbericht* 48: 13–26.

Daszkiewicz, P., Samojlik, T., and Jedrzjewska, B. 2012. *Puszcza Bialowieska w pracach przyrodnikow i podrownikow 1831–1863*. Warsaw: Semper.

Davies, N. 2001. *Heart of Europe: The Past in Poland's Present*. Oxford: Oxford University Press.

Dunn, E. 2004. *Privatizing Poland*. Ithaca, NY: Cornell University Press.

Fleming, M. 2010. *Communism, Nationalism and Ethnicity in Poland. 1944–1950*. New York: Routledge.

Grodzki, W. 2016. "Mass outbreak of spruce bark beetle *Ips typographus* in the context of the Bialowieza Primeval Forest." *Forest Research Papers* 77(4): 324–331.

Grzywacz, A. 2014. "Adam Loret (1884–1940) pierwszy dyrektor naczelny Lasów Państwowych." *Studia i Materiały Ośrodka Kultury Leśnej* 13: 93–120.

Gutowski, J. M. and Krzystofiak, L. 2005. "Directions and intensity of migration of the spruce bark beetle and accompanying species at the border between strict reserves and managed forests in northeastern Poland." *Ecological Questions* 6: 81–92.

Hann, Ch. 1998. "Postsocialist nationalism: rediscovering the past in southeast Poland." *Slavic Review* 57(4): 840–863.

Hann, Ch. 2002. *Postsocialism: Ideals, Ideologies, and Practices in Eurasia*. London: Routledge.

Haraway, D. 2016. *Staying with the Trouble: Making Kin in the Chthulucene*. Durham, NC: Duke University Press.

Harmon, M. and Hua, C. 1991. "Coarse woody debris dynamics in two old growth ecosystems." *Bioscience* 41(9): 604–610.

Humphrey, C. 2002a. *The Unmaking of Soviet Life: Everyday Economies after Socialism*. Ithaca, NY: Cornell University Press.

Humphrey, C. 2002b. "Does the category 'postsocialist' still make sense?," in Ch. Hann (ed.), *Postsocialism: Ideologies, and Practices in Eurasia*, 12–15. London: Routledge.

Jonasova, M. and Prach, K. 2008. "The influence of bark beetles outbreak vs. salvage logging on ground layer vegetation in Central European mountain spruce forests." *Biological Conservation* 141(6): 1525–1535.

Knudsen, I. and Frederiksen, M. 2015. *Ethnographies of Grey Zones in Eastern Europe*. London: Anthem Press.

Konczal, A. 2018. "Trees that must remember." *Eurozine*. www.eurozine.com/trees-must-remember/ (May 22).

Książek, M. 2018. "Szyszko odszedł oni zostali." *Tygodnik Powszechny* 8: 12–15.

Latour, B. 1993. *We Have Never Been Modern*. Cambridge, MA: Harvard University Press.

Matthews, A. 2011. *Instituting Nature. Authority, Expertise, and Power in Mexican Forests*. Cambridge, MA and London: The MIT Press.

Michalski, J., Starzyk, J. R., Kolk, A., and Grodzki, W. 2004. "Zagrożenie świerka przez kornika drukarza—Ips typographus (L.) w drzewostanach Leśnego Kompleksu Promocyjnego 'Puszcza W Grodzki/Leśne Prace Badawcze, 2016, Vol. 77 (4): 324–331 331 Białowieska' w latach 2000–2002." *Leśne Prace Badawcze* 3: 5–30.

Okołów, Czesław. 2009. "History of Protection." *Białowieża National Park: Know it, Understand it, Protect it*. Białowieża, Poland: Białowieża National Park.

Oosthoeak, K. J. and Holzl, R. (eds.) 2018. *Managing Northern Europe's Forests: Histories from the Age of Improvement to the Age of Ecology*. Oxford: Berghahn Books.

Peszkowski, Z. and Zdrojewski, S. (eds.) 2001. *Leśnicy w grobach Katynia*. Warsaw-Łódź-Orchard Lake: Fundacja NOM.

Petrova, S. 2014. *Communities in Transition: Protected Nature and Local People in Eastern and Central Europe*. New York: Taylor and Francis.

Rakowski, T. 2016. *Hunters, Gatherers, and Practitioners of Powerlessness. An Ethnography of the Degraded in Postsocialist Poland*. New York and Oxford: Berghahn.

Robiński, A. 2018. "W lasach rządzi korporacja. Wywiad z prof. Jerzym Szwagrzykiem." *Tygodnik Powszechny* 8: 16–18.

Schama, S. 1995. *Landscape and Memory*. New York: A. A. Knopf.

Scott, J. C. 1998. *Seeing like a State: How Certain Schemes to Improve the Human Condition Have Failed*. New Haven, CT: Yale University Press.

Smart, A. and Smart, J. 2017. *Posthumanism: Anthropological Insights*. Toronto: University of Toronto Press.

Stegners, I. 2010. *Cosmopolitics I*. Minneapolis, MN: University of Minnesota Press.

Szyszko, J. 2015. "A nation without land faces extinction." *Radio Maryja*, December 27, www.radio-maryja.pl/informacje/prof-jan-szyszko-narod-bez-ziemi-jest-narodem-wymarlym/.

Tsing, A. 2015. *The Mushroom at the End of the World: On the Possibility of Life in the Capitalist Ruins*. Princeton, NJ: Princeton University Press.

Verdery, K. 1996. *What was Socialism and What Comes Next?* Princeton, NJ: Princeton University Press.

Verdery, K. 2002. "Whither postsocialism?," in Ch. Hann (ed.), *Postsocialism: Ideals, Ideologies, and Practices in Eurasia*, 12–21. London: Routledge.

Wolfe, C. 2003. *Zoontologies: The Question of the Animal*. Minneapolis, MN: University of Minnesota Press.

Žižek, S. 2001. *Did Somebody Say Totalitarianism? Five Interventions in the (Mis)use of a Notion*. London: Verso.

Video sources

Odsłonięcie pominka Adama Loreta, Strona Lasów Państwowych (2017). www.lasy.gov.pl/pl/informacje/aktualnosci/odslonieto-pomnik-adama-loreta. Accessed: May 22, 2018.

Wideo Wyborcza.pl. Ks. Duszkiewicz: "Martwe drewno to symbol zła i grzechu" (2017). www.facebook.com/wideowyborcza/videos/381119205649656/. Accessed: May 23, 2018.

Part II

On the left

7 Planned environment in a socialist dictatorship

Complex water management and soil improvement in Moravia[1]

Jiří Janáč

On a gray and rainy afternoon in February 1991, the dam of a recently completed reservoir in Nové Mlýny was the site of a peculiar gathering. Two opposing groups of protesters camped on the banks of a large artificial lake at the bottom of a broad valley, until recently covered by one of the last stretches of natural floodplain forests in central Europe. On one side, there was a group of locals and environmental activists who argued that the entire Nové Mlýny project—a system of three shallow reservoirs built over the past two decades—ought to be completely removed and the unique wetland ecosystem in the valley restored to its pre-construction state. Their adversaries defended a full completion of the project, including a construction of large-scale amelioration networks and recreational facilities. These would form a significant part of the overall project whose aim was to transform a swampy and mosquito-infested area into a prosperous agricultural region and a prime destination for mass tourism. Both groups organized petitions and waved banners that clearly spelled out the core of the dispute: while the environmentalists saw Nové Mlýny as a "crime against nature," the project supporters accused them of "ecoterrorism."[2]

The event had broader political ramifications that went beyond the traditional conflict between economic and environmental rationality characterized by antagonism between industrialization/development on the one side and conservation/protectionism on the other. On this day, protesters came to Nové Mlýny to demonstrate their willingness to follow up on the threat they made in the previous autumn in an ultimatum given to the government, symbolically exactly on the first anniversary of the collapse of Communist rule in Czechoslovakia. Conditions of the ultimatum were clearly spelled out: unless the state authorities step in and start a comprehensive re-evaluation of the Nové Mlýny project within the next 100 days, activists would empty the dams by force.[3] Building on a broad social consensus to the effect that wider public participation in political processes is a core principle of a "post-Communist" transformation, the environmental movement articulated its program as a struggle against the legacy of state socialism, a system they identified as the primary cause of the current environmental crisis.[4] They viewed the large-scale system of water reservoirs as an embodiment of the inefficient and bankrupt bureaucracy of the Communist rule, with its ponderous and cumbersome central planning, preference of large-scale solutions, and arrogant approach to nature.[5]

The actors involved had clearly phrased the Nové Mlýny debate as a dispute between democratic and authoritarian rule, whereby the latter was considered intrinsically anti-environmental. A sociological survey conducted among members of both camps suggested that the sources of friction boiled down to a conflict between hydro-bureaucracies (which we call hydrocrats following Molle and Mollinga[6]) that represented the legacy and values of state-socialist dictatorship, and adherents of a biocentric, ecological perspective, who viewed themselves as representatives of the new, liberal-democratic order, with the local population divided between the two camps.[7] Even so, what they demanded was a scientific examination of various possibilities, including the emptying of reservoirs.[8]

Such a picture of an environmental debate in post-socialist countries is hardly new.[9] It fits in with the short-lived radicalization and popularity of environmental movements following the collapse of Communism in Czechoslovakia.[10] What is curious and worth noting, however, is that a narrative that depicts socialist dictatorships as having a disastrous impact on the natural environment still dominates much existing historiographic literature. This perspective, often promoted by ecologists also on an international level had, to some extent, obscured the role which environmental considerations played in socialist planning and in state-socialist practice. In historiography, it was transformed into a familiar hypothesis of "ecocide," which simply declares that unlike the liberal West, countries of the Socialist Bloc did not find a way of effectively coping with environmental challenges brought about by rapid industrialization.[11] This claim implicitly accuses state-socialist regimes of a lack of interest in the environment and emphasizes the "negative" effects of the socialist system on both the environment and environmental movements.[12] It assumes that environmentalism, be it in the form of "ecological modernity," "post-material values," or other movements, developed in state-socialist countries and their societies only toward the end of their existence. In doing so, this claim assumes the validity of Kotkin's thesis about socialist societies being trapped in an unchanging pattern of a Bolshevik (anti-capitalist, anti-individualist) form of industrial modernity.[13]

This chapter has several interrelated goals. First, its aim is to contribute to a revision of the standard narrative, which depicts socialist dictatorships as essentially anti-environmental. Debates of the past two decades had clearly demonstrated that environmental concerns were voiced both in public and political discourse throughout the time of Communist rule,[14] and state socialism actively tried to mitigate the impending ecological crisis on both a global and regional level.[15] Admittedly, however, this process remained extremely technocratic.[16] By tracing the Nové Mlýny debate back to its beginnings in the 1950s, we intend to highlight important continuities in socialist policies on water and environment and place them in a global context. This should help explain how it came about that a project once heralded as the most significant achievement of landscape engineering in central Europe, a massive undertaking developed "in harmony with both the long-term plan of the country's national economic development and the principles of ecology, while duly respecting the ecosystems as well as integrity and conservation of the environment"[17] became a symbol of "totalitarian" crimes against nature,[18] of projects which, if global natural balance was to be re-established, would have to be either "environmentalized" or completely destroyed.

Second, by looking at relations between power and environmental discourse through the lens of the Nové Mlýny project, a unique and massive water management project undertaken by the Czechoslovak Communist government after the hydropower boom of the 1950s, we want to contribute to a debate about expert environmental knowledge and socialist dictatorship, a debate that has been, to a large extent, skewed by predominance of a certain environmentalist perspective.[19] To do that, we need to clarify our understanding of socialist dictatorships. To wit, under the influence of Cold War dichotomies, there is a marked tendency in political literature and especially security studies to interpret the Socialist Bloc as above all "the other" of the liberal West. Nonetheless, social historians have recently pointed out that the socialist project should be understood rather as an alternative form of the Western modernization project, with which it shared both the ideological background in the form of Enlightenment ideals of rationality and development, and the basic premises, which led to a fundamental ambivalence between liberty and discipline.[20] The legitimacy of communist rule rested on modernist promises of social welfare and security. Based on recent research, we can assume that countries of the Socialist Bloc had to cope with, and adapt to, societal challenges similar to those faced by states and societies in the West.[21] This led to the development of a particular form of post-materialist thinking, which may not have been all too apparent in public debates but penetrated the minds of experts and political leaders.[22]

Subsequently, the position of scientists and experts in Communist regimes, who were generally depicted in the literature as mere instruments of power, was contrasted with their role as advisers in Western liberal democracies. While this opposition cannot be dismissed out of hand, it is important to take into account that, while limiting it by ideological considerations, even socialist regimes acknowledged the authority of experts. Furthermore, states of the Socialist Bloc relied, even more than other modernist states, on the authority of science and scientific expertise.[23] The decision to build the Nové Mlýny project therefore should not be interpreted only within the context of the above-mentioned totalitarian narrative, which places all blame on an omnipotent "regime." A careful analysis should view the project as a step forward in the continuous social and technical construction of a socialist dictatorship, albeit a step where environmentalists, together with experts from various other fields, intervened.

This chapter analyzes the environmental debate within a context of technocratic socialist planning through the prism of a single water management project. Throughout the forty years of socialist dictatorship in Czechoslovakia, the Nové Mlýny project reflected the changing ideas, preferences, and inner workings of the socialist state. Literature on the project, written mostly ex-post facto by its environmentalist opponents, suggests that there were four critical moments when the fate and the form of the entire undertaking were challenged. The first came in the 1950s, when the holistic, conservationist ideas associated with Stalin's Great Plan for the Transformation of Nature were compromised by production-oriented goals set by the same Stalinist ideology under a heading of socialist industrialization.[24] The second came in the post-Stalinist environment of the late 1950s and early 1960s, when criticism of Stalinist megalomania, economic productivism, disproportional development, contradictory policies, and centralization combined and led to a conflict between a technocratic, conservationist perspective, which championed reasonable exploitation combined with the protection of natural resources, and protectionism in the sense of a straightforward protection of nature inspired by an emerging post-materialist perspective.[25] The third critical moment came in the early 1970s, when the project was already under construction. At this time, it became the focus of intensive research carried out by local representatives of global environmentalism, yet the public remained largely passive, following the pattern observable in other socialist countries.[26] And finally, in the early 1980s the Nové mlýny debate transformed into a public discourse within a broader context of global ecological movements.[27] This ultimately led to the project's environmentalization, in the 1990s. The following is therefore structured as a drama in four acts.

Act I: seeing like the state

The story of arguably the most significant attempt at planned landscape transformation in postwar Czechoslovakia started back in the 1930s. During the years of the Great Depression, scientists and economists identified a local water regime, characterized by excessive droughts and regular floods during the vegetative season, as a hindrance to modernization. They believed this water regime formed an obstacle to a rational exploitation of the otherwise favorable climatic conditions and fertile soils of the Southern Moravia plain, a lowland area located in the south-eastern corner of the Czech Republic on the confluence of the Morava and Thaya rivers. The dry and hot climate of the early 1930s made the growing discrepancy between available water resources and needs in the whole of Czechoslovakia, but particularly in Southern Moravia, painfully apparent.[28] While the Czech hydraulic bureaucracy called for a centralized water management on a national scale in order to secure a coordinated and efficient water resource development,[29] the agricultural circles worst hit by the weather established a committee whose task was to investigate the causes of the drought, with focus especially on the situation in Southern Moravia.[30]

Even the outbreak of World War II did not manage to push aside these debates on water problems. Quite the contrary: the technocratic system of governance installed by the Nazi regime in the occupied territory of former Czechoslovakia demanded the creation of water resource inventories for all major river basins. Its aim was to secure enough water to satisfy the needs of planned economic development: in this case, the Reich planned a massive inland waterway that would connect the Elbe and Oder with the Danube near Vienna.[31] These efforts resulted in a drafting of a water management plan for the basin of the river Morava, which painted a grim vision of the future. A study of hydrological characteristics of the Thaya river at Dolní Věstonice—on the banks of the current Nové Mlýny dam—clearly showed the extreme fragility of the water balance between needs and resources in the region, especially during the vegetation season. The authors, local water management experts from the university in Brno, assumed that a construction of large reservoirs that would retain water from the spring floods would be needed to secure sufficient amounts of water even during "normal" years (i.e., not drought conditions). Furthermore, after considering the expected and planned intensification of agriculture, industrial development, and growing population density, these experts argued that the Thaya river basin could not provide the required volumes of water, which is why a water transfer from other river basins would be required in the long run.[32] The war economy did not, however, favor such plans—for the time being, therefore, they stayed on paper.

The situation after the war to some extent resembled what happened in the 1930s. Extreme drought in 1947 intensified both the water problems in Southern Moravia as experienced by local farmers and, perhaps even more profoundly, the public and expert awareness of, and interest in, water resource issues. This, in tandem with the local technocratic tradition[33] and the semi-command nature of the postwar reconstruction economy, led to an articulation of a brand new national water resource policy. The last impulse came from politics in the form of a Communist coup d'état in February 1948, largely supported by the Czechoslovak population, which endorsed the Communist program based on a promise of massive modernization for all. This eliminated the legal and institutional obstacles which, up to now, hampered the development of water use and development planning on a national level. In particular, the Communist government proclaimed all water, soil, and water resources a national property and introduced economic planning as a centralized system that should settle disputes between various stakeholders in favor of the "common good," which was, of course, defined as the development of a "communist society."[34]

A detailed account of all available water resources in the country, the State Water Management Plan (Státní vodohospodářský plán, SVP) was developed during the first Communist five-year plan as a background for all economic planning. The plan followed "the logic of nature" and divided the country into three large river basins: Elbe in Bohemia, Morava in Moravia, and Danube in Slovakia. Based on detailed surveys, a crucial part of the plan calculated the present and future water management balance for thirty-five major river basins (the Thaya river basin being among them), thus basically articulating the natural limits of water use for every settlement in the country.[35] The development of this plan was influenced by the traditional modernist technocratic idea of a hydraulic mission, which viewed water as a national resource and emphasized the need for economizing the available resources and utilizing them to the fullest extent possible.[36]

The attitude of the State Water Management Plan toward the natural environment cannot, however, be described as hubristic or anti-environmental. The planners had consulted a wide range of experts and the plan was based on conservationist principles: even the government decree of December 1948, which sanctioned the development of the plan, stated that one of its main goals was to be a restoration of the "biological balance in nature." Indeed, besides providing the basis for resource exploitation and further industrialization, the plan was also explicitly designed so as to correct the negative effects of unregulated, profit-oriented, and insufficiently coordinated capitalist interventions into the water regime inherited from previous decades. This can be largely attributed to the Soviet influence associated with the specific hydraulic ideology of Stalin's Great Plan for the Transformation of Nature. Following the Soviet example, the State Water Management Plan included various measures to reduce "useless runoff' (i.e., loss of water) by implementing a "biological modification of landscape" with the purpose of enlarging its water retention capacity.[37] These measures included the establishment of shelterbelts and ponds, afforestation, and the like.[38]

Regarding Southern Moravia, the State Water Management Plan identified the region as both one of the most arid parts of the country and a crucial agricultural area. To improve the efficiency of local agricultural production, planners proposed to enhance the soil quality by optimization of its hydrological regime. This, naturally, required large-scale amelioration and irrigation in conjunction with the construction of ponds and shelterbelts designed to slow down surface run-off and evaporation. Details of the proposed changes were not specified, but the plan clearly formulated the idea that a "transformation of nature" in Southern Moravia (including the local climate) should be achieved by the alteration of the water regime.[39]

Act II: on the nature of the project of socialist modernity

In 1959, Governmental Decree No. 826 launched the preparation stage of the Comprehensive Water Management and Soil Improvement Scheme for

Southern Moravia.[40] The plan was based on ideas spelled out in the State Water Management Plan (SVP) but also responded to a more general shift in the nature of the Communist rule. The plan got the green light following a dramatic change in Soviet and Czechoslovak water policies characterized by abandoning the hydropower paradigm, with its emphasis on the construction of large, multipurpose dams, which were to be substituted by similarly large land reclamation schemes. While this change is sometimes linked to differences in the personal inclinations of Stalin and Khrushchev,[41] it seems more likely that in the post-Stalinist period, the governments of countries of the Eastern Bloc reoriented their policies in order to avert a looming economic crisis and the associated crisis of legitimacy.

Broadly speaking, the post-Stalinist period was characterized by a less rigid and more critical stance toward communist ideology. The far-reaching transformative aspirations of the Stalinist version of the communist project, associated with intense mobilization of resources (be it water or population), had weakened and made way for a new articulation of route toward socialist modernity. Attempts to substitute Stalin's charismatic leadership with a technocratic administration found its most characteristic expression in the discourse of the Scientific and Technological Revolution which, in a condensed form, offered a synthesis of positivist Soviet ideas about scientific rationalism, formulated in the 1961 program of the Communist Party of the Soviet Union, and a holistic approach to long-term planning.[42] The principal "ideological" component of the Scientific and Technological Revolution with respect to environmental problems was the pervasive conservationist conviction that socialist societies have a unique ability to successfully transform and utilize the natural environment as an economic resource through holistic planning, that is, planning that would consider both nature-related and social aspects of the issue.[43] In effect, this approach was best articulated in the ideas of "landscape transformation," initially developed as an instrument to be used in large-scale long-term spatial planning, following the steps of the traditional Russian landscape science.[44]

In the Czechoslovak context, this process started with a failed attempt at economic reform in 1957 and continued throughout the 1960s. The new constitution of 1960, which proclaimed that the Czechoslovak society had successfully completed a transition from capitalism to socialism, marked in this respect an important turning point. By the late 1950s, it became clear to both experts and the lay public (including politicians) that the attempt at a comprehensive "transformation of nature" had been heavily compromised by the prioritization of economic and political goals and the rapid industrialization had a negative impact on the quality of life, especially by leading to a deterioration of the natural environment.[45] The shift of discourse from a "transformation of nature" toward "landscape protection and formation" was apparent in the re-naming of scientific and administrative institutions in Czechoslovakia in the period of 1957–1960. It testified to a renunciation of the radical revolutionary attitude in favor of the original conservationist ideas, which underpinned Stalin's Great Plan for the Transformation of Nature[46] and to the adoption of premises of the Soviet vision of the Scientific and Technological Revolution that was being formulated at this time.

The first impulse for the realization of the project came from regional administration, in particular from the Regional National Committee of Southern Moravia (Jihomoravský krajský národní výbor), which included the Nové Mlýny reservoirs, as a means of regulating the Morava and Thaya rivers in its regional development plan in 1958. The plan aimed primarily at the intensification of agricultural production in the region by flood protection of forests and fields, irrigation, drainage, conversion of waterlogged meadows into arable land, introduction of more demanding crops, such as corn or rice, and the like. These goals were to be achieved by a transformation of the water regime. This document, prepared by the "planning committee" of the Regional National Committee, took into account the spatial and economic perspective without, however, considering any broader ecological contexts and implications. In order to boost local, so far predominantly rural, economy, it was foreseen that the water held in the reservoir system of Nové Mlýny would be used also by a new thermal power station and a large paper mill and other new industrial facilities.[47]

In the second step, a hydrocratic institution, the Main Office of Water Resources Development (Ředitelství vodohospodářského rozvoje, established in 1960) developed a Prospective Study of Melioration of Morava and Thaya Rivers in Southern Moravia (*Výhledová studie úpravy řek Moravy a Dyje v oblasti jižní Moravy*). This study backed the solution proposed by the Regional National Committee of Southern Moravia. At this time, this Main Office was also working on a new programmatic document concerning the development of water management and water use in Czechoslovakia with an outlook up until the 1980s. The document took as its starting point a premise, formulated by the Scientific and Technological Revolution, about a permanent growth of socialist society and its material welfare, which would lead to ever growing water use both per capita and in total. This plan anticipated a collision of demands. It predicted that unless new dams and other water structures are constructed (plans that would include the pumping upstream of water from the Danube and distributing it across the country by interconnecting all three major river basins), by 1975 Czechoslovak rivers will contain only polluted wastewater.[48] Furthermore, the hydrocrats argued, in such an arid region with an extremely fragile water management balance, the realization of the Nové Mlýny project was the only viable way of addressing and reversing the "underdeveloped character of the region."[49]

The debate than shifted to a scientific ground, when the Czechoslovak Academy of Sciences organized a meeting on *The Impact of Construction of Water Structures on Vegetation* (Vegetační problémy budování vodních děl), a conference dedicated to this issue. At this point, in the spring of 1963, almost all experts involved had reached a consensus to the effect that the hydrological regime of Southern Moravia needs to be altered (and improved). On the other hand, experts from different fields could not agree on the character of the intervention and methods to be used. Forestry experts, concerned with degradation of the floodplain forest in the area, which resulted from a changed flooding pattern in the last decade, advocated instead of a large dam, a small-scale solution based on a restoration of drainage systems built in the nineteenth and early twentieth century. Their main concern was to re-establish a balance and save these forests. They supported their arguments by pointing to economic losses associated with a low annual increment of timber.[50]

Institutions representing the emerging perspective of landscape ecology inspired by the above-mentioned transition in the articulation of the route toward socialist modernity, challenged the plan to build the dams from the very outset. At the above-mentioned debate, as well as in the evaluation process of the Prospective Study, representatives of biological and agricultural committees of the Czechoslovak Academy of Sciences were pointing out that the project did not take into account any of the larger issues of landscape biology. They argued, for instance, that one cannot safely assume that waterlogged meadows could be easily turned into arable land; that a dam construction would most certainly have a negative impact on the remaining stands of floodplain forests; that a shallow dam at Nové Mlýny would provide a perfect environment for malaria-transmitting mosquitoes; and that the suggested water regime alterations would inevitably negatively affect various protected landscapes elsewhere in the region.[51]

In effect, the debate took the shape of a clash between biologists (specialists in forestry and ecology) and water management experts or hydrocrats. One of the more visible manifestations of these tensions came in 1965, when *Vesmír*, a popular science magazine devoted to natural sciences and published by the Academy of Sciences, ran a picture of the Nové Mlýny area with a caption stating that biologists warn against this possibly harmful and hubristic project which may turn into a disaster.[52] Hydrocrats responded in a rather peculiar way: instead of publishing a counterstatement, they sent a complaint to the Department of Agriculture of the Regional National Committee of Southern Moravia, which turned to another branch of the Academy of Sciences, namely the Committee for Water Management, for a response.[53] The then head of the committee, Theodor Ježdík, one of the leading dam builders in Czechoslovakia, answered politely but accused the ecologists that rather than participate in the endeavor, they chose to criticize all the ongoing efforts aimed at soil amelioration and water management. Instead, Ježdík argued, they should provide assistance to engineers whenever needed, and reserve their recommendations and fears for an expert debate, where they would not alarm the public.[54]

The central question of the debate was thus essentially environmental: how to integrate a technological structure developed for agricultural purposes, namely irrigation (the originally envisioned powerplant had disappeared from the debate in the early 1960s), within the ecology of a floodplain in the least harmful way. In other words, how to design and create a controlled and stable natural environment that would enable economically effective utilization and would not constrain the economic development of the region.

Act III: seeing like the world? The arrival of landscape engineering

> For landscape ecologists, Nové Mlýny is the perfect ground for experimentation. And that, in my opinion, is the only benefit of the project.
> (Antonín Buček, landscape ecologist[55])

By the late 1960s, it became clear that the project would be constructed. In addition to the need to avert an impending water shortage, there also emerged in the mid-1960s a policy of economic autarchy in agricultural production, which was to become a new and decisive argument in favor of the project's realization.[56] The protests and objections of biologists and foresters, voiced regularly in specialized journals and at conferences dedicated to the development of the Nové Mlýny project throughout the 1960s, were directed against both the general approach adopted by the planners, namely the construction of large reservoirs, and against the framework of the debate, which was dominated by hydrocrats, agricultural engineers, and economic planners.[57] In the mid-1960s, the central planning institutions considered the project to be fully "sorted out" and ready for implementation. And it was ready, but only from the water management perspective, that is, in terms of providing a technological solution for river regulation and irrigation in the area.

Ecological tendencies, which had been represented in the holistic approach to landscape planning and landscape creation, were not, however, fully considered and integrated. An evaluation of the final form of technological project, conducted by the Geographical Institute of the Academy of Science in 1967, echoed and summarized many of the earlier objections. Under the heading of a "geo-ecological" prognosis, scientists from various fields concluded that, given the unique value of wetland ecosystems in the area, the envisaged economic benefits seemed inadequate and, considering the dubious premises underlying their calculation—such as ecologically unsound assumptions about the possibility of turning river terraces into arable land—even questionable.[58]

The critique used the notion of a biosphere, which emphasized the crucial role of water, wetlands, and floodplain forests in the "wider scheme of things," that is, within the various ecological networks that underpin the stability of the natural environment. This wording reflected a specific Soviet or socialist understanding of the global environmental crisis, which was based on a revival of the original Leninist Communist holistic conservationism in the USSR. This approach was represented mainly by a return to Vernadsky's theses about biosphere and noosphere in the early 1960s.[59] With respect to water issues, instead of emphasizing the traditional, nationally framed narrow definition of hydrological cycle, biosphere-driven thinking stressed the wider and more comprehensive relations between water and living organisms.[60]

The concept of a biosphere became even more prominent in the new stage of development of the Nové Mlýny project, which was significantly influenced by globalization of the Czechoslovak ecological debate. While the geo-ecological prognosis did not shake the confidence of hydrocratic circles, the state authorities took into account the wider (and not only) ecological dimensions of the project. This resulted in the creation of an interdepartmental government committee, whose task it was to secure coordination and harmonization of various perspectives in the realization of the project. Construction was launched in 1968 with the regulation of the Thaya downstream from the planned reservoir and the first dam, that is, in the least disputed sections of the entire scheme.[61] Preparations for the construction of the other two reservoirs were to be subjected to a further inquiry based on experiences gained in the course of the realization of the first phase. The environmental impact of water regime alteration was the main subject of research that formed part of the Czechoslovak contribution to the UNESCO Man and the Biosphere project (MAB), one of the very first examples of global environmentalism. This research was conducted by an MAB project coordination center at the Brno Agricultural University.[62]

The importance of UN science programs for the development of global scientific networks and thereby also for global environmentalism has been noted by contemporaries and recently also by historians. Their role in eastern Europe, however, has not yet received sufficient attention. In the Czechoslovak case, UN programs such as the International Geophysical Year, but especially the International Biological Programme and the subsequent project Man and the Biosphere, led to a radical revision of science policies and ultimately also affected decision-making in other areas, thus in effect strengthening the position of scientists as experts.[63] By the late 1960s, a university institute dedicated specifically to the agenda of the International Biological Programme was established at the Brno Agricultural University. It was headed by Professor Miroslav Penka and later evolved into an Institute for International Biological Programme.

In effect, early 1970 witnessed both the beginning of the realization of the Nové Mlýny scheme and its gradual transformation from an instrument of primary water management, aimed at increasing agricultural productivity in the region, into an ecologically informed and comprehensive project of landscape transformation. In 1973–1974, the government committee organized a sort of open competition aimed at developing the most comprehensive approach possible: seven teams of experts analyzed the project from various perspectives, ranging all the way from urban/rural connections to landscape planning and even futurology.[64] At the same time, a general project for the transformation of Southern Moravia was being developed under the title "A Comprehensive Project of Landscape Care with Respect to Water Management Measures in Southern Moravia" (*Generel péče o krajinu v oblasti vodohospodářských úprav jižní Moravy*) by TERPLAN, the Czechoslovak State Institute for Territorial Planning. The document built on knowledge gathered by other teams and focused on achieving the long-term ecological stability of the new landscape.[65]

By the mid-1970s, the Southern Moravian experiment reached a point where the authorities contemplated proclaiming the territory, then undergoing a vast transformation, a "model area of comprehensive landscaping." This would imply protection resembling the regulated preservation regime of national parks. Its aim would be the creation of a protected environment suitable for studying the possibilities of "engineering the landscape" by harmonizing economic planning with ecological concerns. Southern Moravia would become a laboratory for testing cooperation between technical and biological methods, the two main approaches that have the potential to solve the apparent conflict between the concerns of ecology (care for the landscape) and economy (use of its natural resources).[66]

Act IV: ecocide

Disagreements between experts, especially between biologists on the one hand and agricultural and hydraulic engineers on the other, were voiced in public. In 1976, after the opening of the first dam, the national journal on nature protection *Památky a příroda* published several articles that criticized the effort to radically alter the landscape. Although these critics admitted that Nové Mlýny is "so far the best planned water control structure" in Czechoslovakia, they still objected to the fact that economists got the upper hand and that the resulting disappearance of a floodplain forest and habitat cannot be redeemed by any (questionable) economic profit. This was illustrated in the example of transformation of the Thaya riverbed: while naturalists and biologists urged for a wide riverbed with levees that would be compatible with a survival of the forest, engineers demanded a flow constriction by a narrow riverbed that would protect from flooding as much land as possible. While the result was a sort of compromise (the left bank was designed to allow regular flooding), it nonetheless condemned the floodplain forest to death.[67]

In the years that followed, almost every issue of various conservationist journals of the time included some article on the disastrous effects of the loss of the South Moravian floodplain. This criticism was fueled by three interconnected elements: (1) recurring ecological problems on the construction site (fish and bird toxification, dam erosion, etc.); (2) the fact that the project was implemented in a "stripped down" form, with the result that what was being constructed was basically the 1965 water management scheme that ignored most of the ecological (and other) concerns; (3) the wide ongoing research (MAB). Most of these papers started with a ceremonial praise of the (envisaged) positive effects and indispensability of the project, then voiced certain doubts regarding the actual impact, and finally proceeded to openly criticize the unquestionable fact that the floodplain ecosystem will disappear, forever, and "together with many priceless and unique places that are the home of indigenous forms of our flora and fauna."[68]

By the early 1980s, critical voices were heard even beyond the academic circles, and the Nové Mlýny project entered a new phase that was characterized by a growing public involvement. It should be noted that, while postwar Sovietization had disrupted the continuity of bottom-up protectionism in Czechoslovakia, since the late 1950s, a movement of naturalists (protectionists rather than conservationists) led by Otakar Leiský, a professional zoologist, started to establish itself within an institutional setting provided by the National Museum in Prague. In 1969, this movement evolved into an independent structure known under the abbreviation "TIS." The era of normalization, which arrived in the 1970s, gradually halted this development. In 1979, this formerly grass-roots movement, driven by a protectionist ideology, was forcibly transformed into a top-down nation-wide organization, a formally non-governmental Czechoslovak Union for Nature Conservation (ČSOP). Its stated aim was to function as a platform for public interest in environmental issues and to improve and open planning procedures so as to include public participation on a local level.[69]

This was not, however, the only expression of a growing environmental awareness. Another movement was initiated by scientists from the Institute for Landscape Ecology of the Academy of Sciences, who launched in 1974 an initially one-year educational campaign inspired by the UN Conference on Human Environment in Stockholm. This initiative, supported by the Czech Committee of the Socialist Union of Youth (associated with the Communist Party of Czechoslovakia), as part of a "year of environment protection and creation," took as its symbol a brontosaurus because it "outgrew its limits." The Brontosaurus movement quickly grew and organized camps and campaigns (known as "akce," i.e., actions) dedicated to a protection and/or restoration of concrete historical or natural monuments.[70] The floodplain that was undergoing a conversion into Nové Mlýny reservoirs became one of such sites when the construction of the third dam was launched.

Akce dno (Operation Bottom) was an implicit protest event organized mainly by scientists who operated either within the Brontosaurus movement or within the Czechoslovak Union for Nature Conservation. It aimed at rescuing the summer snowflake (*Leucojum aestivum*) growth from an area that would become the bottom of the third Nové Mlýny dam. In charge of this undertaking was Antonín Buček, a scientist employed at the Institute of

Geography of the Czechoslovak Academy of Sciences. Using his position as head of the Czechoslovak Union for Nature Conservation and the former TIS organization in Brno, in 1981 to 1988 he organized several voluntary replanting campaigns. These became quite popular with students of the Brno University, especially after 1984, when an announcement of yet another campaign was published in *Naší přírodou*, the official journal of the Czechoslovak Union for Nature Conservation. Between 1983 and 1988, over 1,100 volunteers joined in the effort to replant the summer snowflake. Most were from Brno, but volunteers arrived from all over Czechoslovakia.[71]

Events organized as part of *Akce dno* fit squarely within the agenda of a growing environmental movement in Czechoslovakia, which attracted an increasing number of people especially after 1984, when a secret report on the state of the environment in Czechoslovakia leaked to the Western media (and through them back to Czechoslovakia). In the West, this inspired a discussion about "eco-bankruptcy" (or "ecocide").[72] Hydrocratic projects and initiatives become symbols of the environmental crisis and the public discourse placed the blame for this on the technocratic management and megalomania. It seemed that the time of massive landscape-altering projects was over.

Environmental policies linked to Soviet Perestroika marked the end of the famous Siberian diversion project in 1987[73] and, on a smaller and more local scale, in February 1988, a civic petition against a hydroelectric dam, Křivoklát, put an end to that project.[74] The opening part of this chapter describes a similar story: in early 1990, there even came an initiative aimed at emptying the Nové Mlýny reservoir in order to save the remaining and still undamaged floodplain ecosystems. In a word play on the former *Akce dno* (Operation Bottom) and in order to highlight the continuity of efforts to limit the negative impact of the Nové Mlýny project, this initiative was called *Akce špunt* (Operation Drain Stopper).

Conclusion

The Comprehensive Water Management and Soil Improvement Scheme in Southern Moravia aimed at a complex landscape transformation driven by economic concerns: it focused on large-scale irrigation, forestry, and control of the water regime of the Thaya River. Its authors and advocates were well aware that to ensure a successful realization of the project, it was essential to involve and integrate a wide range of expert knowledge from other fields, including ecology. Since the very beginning of Sovietization in Czechoslovakia after 1948, the state-socialist dictatorship aimed at a "harmonious development" of society and nature (human environment). This was underpinned by an ideological claim to the effect that a socialist society and socialist state are in a unique position to deal with ecological challenges. They are in this position thanks to adopting a scientific and rational (plan-based) approach to the utilization of both natural resources and the development of society, which aimed at sustainability and balance instead of individual profit. And while this approach was repeatedly and regularly compromised by various urgent and pressing political preferences dictated by the need to sustain the Communist order and its legitimacy (socialist industrialization, agricultural autarky, water shortage, the economic crisis of the 1980s), it never disappeared from the sight of socialist leaders altogether. This is perhaps best illustrated by the concept of landscape ecology. Already the first socialist Nature Protection Act of 1956 operated with the notion of landscape as a subject of protection and this holistic concept gradually, and undoubtedly due to Soviet influence, became a central instrument of territorial planning.

In the Czechoslovak context, the actors involved translated the ongoing dispute between the immediate needs and long-term planning into a paradigmatic dispute about the value of nature.[75] This discussion was echoed in the clash between the hydrocrats and ecologists mentioned in the opening paragraphs. While engineers preferred to build and deliver, despite the threat of environmental losses resulting from a shortage of investment funds, gaps in planning methodology, and general delays in planning, environmentalists and conservationists opposed such ambitions and tried to halt the realization of massive, yet insufficiently researched, interventions into the fragile ecological balance. This dispute can be clearly identified already in the discussions surrounding Stalin's Great Plan for the Transformation of Nature. Despite a shared belief in technology's potential to transform nature, when hydrocrats pursued their building agenda even without the additional "environmental/non-productive" investments, conflicts came to the fore. Over time, the shift from a traditional, holistic, and conservationist approach to water toward an openly sectoral water management, where water resource development acts as if on demand from other branches of economy, exacerbated the dispute. By the end of the 1980s, the opposition was framed differently: a sort of post-materialist environmentalism aligned with a perspective of human security was contrasted with a state-centered concept of environmental security, which viewed the environment still above all as an important source of national economy and, following technocratic rationality, advanced the "collective" interests in the form of "large-scale" solutions.[76]

Attempts to reconcile the two paradigms were—and this seems important as far as the nature of state-socialist environmentalism is concerned—apparently meant seriously but remained futile, which was at least in part due to the hopelessly technocratic character of state-socialist environmentalism. The landscape ecology approach provided a useful, but again expert-led and in its own way, technocratic platform for discussion. The Nové Mlýny project was in effect developed on a relatively broad platform of actors, but inhabitants of the village of Mušov, now on the bottom of the Nové Mlýny reservoir, were for the first time formally consulted (or, to be precise, informed) about the planned dam in 1967, yet even so, the resettlement of Mušov was uncontested until the village was actually flooded in 1978.[77]

Vlastimil Vaníček, an agricultural and environmental expert and an advocate of Nové Mlýny construction in the 1970s and 1980s, is a perfect specimen of such a "technocratic environmentalist." As a regional biologist (this was a position in regional administration), in 1968 he called the biologists and geographers from the Academy of Sciences who opposed the project "folklore collectors" and foes of the new socialist landscape.[78] And yet, he also wrote extensively about the need to improve the rural environment.[79] In the end, he became a central figure in the MAB project. But the technocratic character of socialist environmentalism is perhaps best illustrated by the nature of the dam-emptying debate of the 1990s, in which opponents of the project called explicitly for an "expert evaluation," thus evincing a technocratic attitude whose roots stretched back to the Stalinist 1950s.

Notes

1 This research was supported by the Grant Agency of the Czech Republic, project GAČR 18-05095Y.

2 Březina, I. (2009). *Zelená apokalypsa: průvodce eko-strachem přelomu milénia*. Praha, Centrum pro ekonomiku a politiku.

3 Kundrata, M. (1991). "Kalendář 100 dní kolem Nových Mlýnů," *Veronika* 6, no. 2: 39–40.

4 Vaněk, M. (2002). Ostrůvky svobody: kulturní a občanské aktivity mladé generace v 80. letech v Československu. Praha, Ústav pro soudobé dějiny AV ČR.

5 Views presented in interviews shot during the gathering in February 1991. Akce Špunt, documentary film, Sky Productions, 1991.

6 Molle, François and Mollinga, Peter P. (2009). "Hydraulic Bureaucracies and the Hydraulic Mission: Flows of Water, Flows of Power," *Water Alternatives* 2, no. 3: 328–349.

7 Kundrata, M. and Musil, L. (1992). "Novomlýnský konsensus?" *Veronika* 7, no. 2: 1–5.

8 Šebela, M. (2005). Živá voda pod Pálavou: zločin pod Pálavou. Brno, Moravské zemské muzeum.

9 Galambos, J. (1993). "An International Environmental Conflict on the Danube: The Gabcikovo-Nagymaros Dams," in A. Vari and P. Tamas (eds.), *Environment and Democratic Transition*. Technology, Risk, and Society (An International Series in Risk Analysis), vol. 7. Dordrecht: Springer.

10 Fagan, Adam (2004). *Environment and Democracy in the Czech Republic: The Environmental Movement in the Transition Process*. Cheltenham: Edward Elgar; Dawson, J. I. (1996). *Eco-nationalism: Anti-nuclear Activism and National Identity in Russia, Lithuania, and Ukraine*. Durham, NC: Duke University Press.

11 Dawson, J. I. (1996). *Eco-nationalism: Anti-nuclear Activism and National Identity in Russia, Lithuania, and Ukraine*. Durham, NC: Duke University Press; Josephson, Paul R. (2013). *An Environmental History of Russia*. Cambridge: Cambridge University Press; Uekötter, Frank (2013). "Environmentalism, Eastern European Style. 'Some Exploratory Remarks,'" in *Umweltgeschichte(n): Ostmitteleuropa von der Industrialisierung bis zum Postsozialismus; Vortrage der Tagung des Collegium-Carolinum in Bad Wiesseevom 4. bis 7. November 2010*. Göttingen: Vandenhoeck

& Ruprecht, 241–254; Weiner, Douglas R. (1999). *A Little Corner of Freedom. Russian Nature Protection from Stalin to Gorbachev*. Berkeley, CA: University of California Press.

12 Hadler, Markus (2017). *Influence of Global Ideas on Environmentalism and Human Rights: World Society and the Individual*. London: Palgrave Macmillan, 99.

13 Kotkin, Stephen (1995). *Magnetic Mountain: Stalinism as Civilization*. Berkeley, CA: University of California Press.

14 Huff, Tobias (2015). *Natur und Industrie im Sozialismus. Eine Umweltgeschichte der DDR*. Göttingen: Vandenhoeck & Ruprecht; Obertreis, Julia (2012). "Von der Naturbeherrschung zum Okozid? Aktuelle Fragen einer Umweltzeitgeschichte Ost- und Ostmitteleuropas," *Zeithistorische Forschungen* 9, no. 1: 115–122; Štanzel, Arnošt (2017). *Wasserträume und Wasserräume im Staatssozialismus: Ein umwelthistorischer Vergleich anhand der tschechoslowakischen und rumänischen Wasserwirtschaft 1948–1989*. Göttingen: Vandenhoeck & Ruprecht.

15 Brain, Stephen (2010). "The Great Stalin Plan for the Transformation of Nature," *Environmental History* 15, no. 4: 670–700; Olšáková, Doubravka (ed.) (2016). *"In the Name of the Great Work": Stalin's Plan for the Transformation of Nature and its Implementation in Eastern Europe*. Oxford: Berghahn.

16 Spurný, Matěj, Olšáková, Doubravka, Sommer, Vítězslav, and Janáč, Jiří (2017). "Technokratischer Sozialismus in der Tschechoslowakei: was ist technokratischer Sozialismus? Bohemia: Zeitschrift für Geschichte und Kultur der böhmischen Länder," *A Journal of History and Civilisation in East Central Europe* 57, no. 1: 12–24.

17 Vaníček, Vlastimil and Hrabal, Antonín (1974). "The Engineering Landscape in a Historical Area of Central Europe," *Landscape Planning* 1, no. 1: 57–79.

18 Šebela, M. (2005). *Živá voda pod Pálavou: zločin pod Pálavou*. Brno, Moravské zemské muzeum.

19 Josephson, Paul R. (1996). *Totalitarian Science and Technology*. Atlantic Highlands, NJ: Humanities Press; Josephson, Paul R. (2002). *Industrialized Nature: Brute Force Technology and the Transformation of the Natural World*. Washington, DC: Island Press/Shearwater Books.

20 Delanty, G. (2013). *Formations of European Modernity: A Historical and Political Sociology of Europe*. Basingstoke: Palgrave Macmillan; Wagner, P. (1993). *A Sociology of Modernity: Liberty and Discipline*. London: Routledge.

21 Krylova, Anna. (2014). "Soviet Modernity: Stephen Kotkin and the Bolshevik Predicament," *Contemporary European History* 23, no. 2: 167–192.

22 Huff, Tobias (2015). *Natur und Industrie im Sozialismus. Eine Umweltgeschichte der DDR*. Göttingen: Vandenhoeck & Ruprecht; Olšáková, D. and Janáč, J. (2018). *Kult jednoty: stalinský plán přetvoření přírody v Československu 1948–1964*. Praha: Academia; Spurný, M. (2016). *Most do budoucnosti: laboratoř socialistické modernity na severu Čech*. Praha: Karolinum.

23 Olšáková, D. (2017). "Im globalen Netzwerk? Die Planung der wissenschaftlich-technischen Entwicklung der Tschechoslowakei von der Sowjetisierung bis zu den ersten Integrationsversuchen in den sozialistischen Block," *Bohemia* 57, no. 1.

24 Gestwa, Klaus (2010). *Die Stalinschen Grossbauten des Kommunismus: sowjetische Technik- und Umweltgeschichte, 1948–1967*. München: R. Oldenbourg.

25 Breyfogle, Nicholas B. (2015). "At the Watershed: 1958 and the Beginnings of Lake Baikal Environmentalism," *The Slavonic and East European Review* 93, no. 1: 147–180.

26 Pál, Viktor (2011). "To Act or not to Act: Water Problems in North-Eastern Hungary after 1945," in T. Myllyntaus (ed.), *Thinking through the Environment*. Cambridge: The White Horse Press, 268–288.

27 Uekötter, F. (2010). "The End of the Cold War. A Turning Point in Environmental History?" in J. R. McNeill and C. Unger (eds.), *Environmental Histories of the Cold War*. Cambridge: Cambridge University Press, 343–351.
28 Úlehla, Vladimír (1947). *Napojme prameny*. Život a práce: Praha.
29 Janáč, Jiří (2017). Hydrokraté a česká vodohospodářská mise ve dvacátém století (1930–1960). *Soudobé dějiny* 25, no. 1: 219–253.
30 Úlehla, Vladimír (1947). *Napojme prameny*. Život a práce: Praha.
31 Janáč, Jiří (2012). *European Coasts of Bohemia: Negotiating the Danube-Oder-Elbe Canal in a Troubled Twentieth Century*. Amsterdam: Amsterdam University Press.
32 Bažant, Jan (1941). Moravský vodohospodářský plán, Povšechná studie. Brno.
33 Elisabeth van Meer, (2010). " 'The Nation Is Technological': Technical Expertise and National Competition in the Bohemian Lands, 1800–1914," in M. Kohlraush and K. Steffen (eds.), *Expert Cultures in Central Eastern Europe: The Internationalization of Knowledge and the Transformation of Nation States since World War I*. Osnabrück: Fibre, 85–104.
34 Dub, Oto (1953). *Vodné hospodárstvo za kapitalismu a socialismu*. Martin: Osveta; Vitha, Oldřich (1952). "Veliké stavby komunismu a rozvoj vědy," *Sovětská věda—stavebnictví* 1, no. 1: 6–9.
35 Státní vodohospodářský plán republiky Československé (SVP, 1954), ÚSVH: Praha.
36 Molle, François and Mollinga, Peter P. (2009). "Hydraulic Bureaucracies and the Hydraulic Mission: Flows of Water, Flows of Power," *Water Alternatives* 2, no. 3: 328–349.
37 Státní vodohospodářský plán republiky Československé (SVP, 1954), ÚSVH: Praha.
38 Olšáková, D. and Štanzel, A. (2016). "Kafkasque Paradigms," in D. Olšáková (ed.), *In the Name of the Great Work: Stalin's Plan for the Transformation of Nature and its Impact in Eastern Europe*. New York and Oxford: Berghahn.
39 Státní vodohospodářský plán republiky Československé (SVP, 1954), ÚSVH: Praha.
40 Pavlík, Stanislav and Hrabal, Antonín (1983). *Vodohospodářská výstavba Jižní Moravy*. Praha: SZN.
41 Josephson, Paul R. (1995). " 'Projects of the Century' in Soviet History: Large-Scale Technologies from Lenin to Gorbachev," *Technology and Culture* 36, no. 3: 519–559.
42 Guth, Stefan (2015). "One Future Only. The Soviet Union in the Age of the Scientific–Technical Revolution," *Journal of Modern European History* 13, no. 3: 355–376.
43 Tupytsia, J. (1977). "Nature Conservation as a Problem of Economic Science," *Problems in Economics* 21, no. 8: 17–23.
44 Shaw, D. J. B. and Oldfield, J. (2007). "Landscape Science: A Russian Geographical Tradition," *Annals of the Association of American Geographers* 97, no. 1: 111–126.
45 Málek, Ivan (1960). "Úvodní slovo," in J. Jeník (ed.), *Tvorba a ochrana Krajiny: Sborník statí projednaných na 6. plenárnim zasedáni biologicko-lékařaké sekce* ČSAV *dne 29. řijna 1959 a zasedáni komise pro ochranu přírody a tvorbu krajiny dne 19. listopadu 1959*. Praha: ČSAV.
46 Brain, Stephen (2010). "The Great Stalin Plan for the Transformation of Nature," *Environmental History* 15, no. 4: 670–700; Olšáková, D. and Janáč, J. (2018). *Kult jednoty: Stalinský plán přetvoření přírody v Československu 1948–1964*. Praha: Academia.
47 Horák, J. (1991). "K historii jednoho vodního díla," *Veronika* 5, no. 1: 33–37.

48 Vitha, Oldřich (1965). "Perspektivní vodohospodářská bilance ČSSR," *Plánované hospodářství* 18, no. 5: 37–43.
49 Výhledová studie úpravy řek Moravy a Dyje v oblasti jižní Moravy, appending to the technical and economic justification of alterations, 1962, p. 50. Archive of the Czech Academy of Science, Prague (AAVČR), archival group Watermanagement Committee of the Czechoslovak Academy of Sciences 1954–1970/Komise pro vodní hospodářství ČSAV (KVH), box. 27.
50 Skatula, Leo (1961). "Studie podmínek pro úpravu vodního režimu oblasti lužních lesů na jižní Moravě, 78–91," in *Vegetační otázky při budovaní vodních děl*. Praha: manuscript. AAVČR, f. KVH ČSAV, b. 34.
51 Ochrana přírody 18, no. 4, 1963—the journal for nature protection published by the Czechoslovak Academy of Sciences devoted a special issue to the critical evaluation of the planned water regime alterations in Southern Moravia.
52 Leiský, Otakar and Bílek, Bohumil (1965). "Čas v programu biologů," *Vesmír* 44, no. 3: 74–86.
53 Koryta, Jiří (1988). "Velké stavby kolem nás," *Vesmír* 67, no. 6: 326–327.
54 Ježdík, Theodor (1966). "K času v programu biologů," *Vesmír* 44, no. 1: 22.
55 Buček, Antonín (2007). "Krajina novomlýnských nádrží jako geoekologický experiment," in V. Herber (ed.), *Fyzickogeografický sborník 4*. Brno: Masaryk University, 81–86, 81.
56 Orsillo, Nicholas (2013). "The Socioeconomic Factors Behind Agricultural Land Drainage's Environmental Impact in Communist-Era Czechia," in *Umweltgeschichte(n): Ostmitteleuropa von der Industrialisierung bis zum Postsozialismus; Vortrage der Tagung des Collegium Carolinum in Bad Wiessee vom 4. bis 7. November 2010*. Göttingen: Vandenhoeck & Ruprecht, 209–228.
57 Horák, J. (1991). "K historii jednoho vodního díla," *Veronika* 5, no. 1: 33–37.
58 Demek, J. a kol. (1967). *Komplexní geografický výzkum zájmového území přehrady na Dyji u Nových Mlýnů*. Brno: ČSAV.
59 Bailes, K. (1981). "Science, Philosophy and Politics in Soviet History: The Case of Vladimir Vernadskii," *The Russian Review* 40, no. 3: 278–299.
60 Budyko, Mikhail Ivanovich (1984). *Evolyutsia Biosfery*. Moscow: Gidrometeoizdat; Hasík, Otakar (1974). *Vodohospodářská výstavba a životní prostředí člověka*. Praha: Academia.
61 Buček, Antonín (2007). "Krajina novomlýnských nádrží jako geoekologický experiment," in V. Herber (ed.), *Fyzickogeografický sborník 4*. Brno: Masaryk University, 81–86; Pavlík, Stanislav and Hrabal, Antonín (1983). *Vodohospodářská výstavba Jižní Moravy*. Praha: SZN.
62 Penka, Miroslav (1991). *After Water Management Measures*. Amsterdam: Elsevier.
63 Olšáková, D. (2017). "Im globalen Netzwerk? Die Planung der wissenschaftlich-technischen Entwicklung der Tschechoslowakei von der Sowjetisierung bis zu den ersten Integrationsversuchen in den sozialistischen Block," *Bohemia* 57, no. 1.
64 Vaníček, Vlastimil and Antonín, Hrabal (1974). "The Engineering Landscape in a Historical Area of Central Europe," *Landscape Planning* 1, no. 1: 57–79.
65 Petříček, Václav (2007). *Významné krajinné prvky včera, dnes a zítra. Vzpomínka na Igora Míchala, zemřel před pěti lety. Sborník z konference: "ÚSES—zelená páteř krajiny."* Brno: Masarykova univerzita, 1–9.
66 Vaníček, Vlastimil and Antonín Hrabal (1974). "The Engineering Landscape in a Historical Area of Central Europe," *Landscape Planning* 1, no. 1: 57–79.
67 Musil, Oldřich (1976). "Jihomoravská balance," *Památky a příroda (Ochrana přírody)* 31, no. 5: 495–500.
68 Adámek, Zdeněk (1976). "Bentos zaplavovaných jihomoravských luk," *Památky a příroda (Ochrana přírody)* 31, no. 4: 251–253, 252.

69 Snajdr, Edward (2008). *Nature Protests: The End of Ecology in Slovakia*. Seattle, WA: University of Washington Press; Vaněk, M. (1996). *Nedalo se tady dýchat. Ekologie v českých zemích v letech 1968–1989*. Praha: Maxdorf.
70 Snajdr, Edward (2008). *Nature Protests: The End of Ecology in Slovakia*. Seattle, WA: University of Washington Press; Vaněk, M. (1996). *Nedalo se tady dýchat. Ekologie v českých zemích v letech 1968–1989*. Praha: Maxdorf.
71 Buček, Antonín (2011). "Akce Dno," *Veronika* 25, no. 6: 2–3.
72 Vaněk, M. (1996). *Nedalo se tady dýchat. Ekologie v českých zemích v letech 1968–1989*. Praha: Maxdorf, 68–71; Feshbach, Murray and Friendly, Alfred (1992). *Ecocide in the USSR. Health and Nature under Siege*. New York: Basic Books.
73 Duke, D. F. (2006). "Seizing Favours from Nature: The Rise and Fall of Siberian River Diversion," in T. Tvedt and E. Jakobsson (eds.), *A History of Water, Volume 1: Water Control and River Biographies*. London: I. B. Tauris, 3–35.
74 Vaněk, M. (1996). *Nedalo se tady dýchat. Ekologie v českých zemích v letech 1968–1989*. Praha: Maxdorf.
75 Huba, M. (2017). "Z histórie paradigmatického sporu o budúcnosť přírody a krajiny Podunajska," *Forum Historiae* 11, no. 1: 141–161.
76 Koryta, Jiří (1988). "Velké stavby kolem nás," *Vesmír* 67, no. 6: 326–327.
77 Čapka František (2000). "Poslední dějství Mušova," in E. Kordiovský (ed.), *Mušov 1278–2000*. Znojmo: FPO Znojmo, 361–388.
78 Horák, J. (1991). "K historii jednoho vodního díla," *Veronika* 5, no. 1: 33–37.
79 Vaníček V. (1968). Význam meliorací v tvorbě a ochraně životního prostředí. Výstavba socialistické vesnice 3/1968.

8 Slovakia's economic and political development in the communist regime of post-1948 Czechoslovakia and its environmental context

Ľudovít Hallon and Miroslav Sabol

Introduction

In 1945, the territory of the current Slovak republic became an integral part of a restored Czechoslovakia, as its economically less developed eastern component. After the installation of a communist regime and subsequent establishment of central economic planning in 1948, Slovakian leaders were confronted with several problems. These included the ongoing centralization of the totalitarian state, where the key decision-making governmental institutions were outside the Slovak territory, in Prague. At the same time, the backwardness of the Slovak economy (in comparison with its Czech counterpart) was targeted for development via intensive state-directed industrialization, with a severe impact on the environment.

The environmental consequences of socialist industrialization in central and eastern Europe have been the subject of numerous studies at the national and global scales. Specific environmental issues related to the socialist industrialization of Slovakia have been studied mostly by Slovakian natural scientists, such as M. Huba, a geographer and recognized personality of the Slovak environmental movement.[1] The topic of ecological monitoring has been of marginal interest in Slovak historiography for a long time. For example, a monograph published by the Slovak historian J. Mlynárik, who performed on Free Europe radio in the 1980s, belongs to a very small set of studies dedicated to this matter.[2] The study of the Slovak historian M. Barnovský, *The Industrialization of Slovakia and Environment in the Communist Era*, dated in 2007 and based mainly on original archival sources, represented a milestone in historical research and the development of the environmental protection topic in Slovakia during the period from 1948 to 1989.[3] The following study is a follow-up of the pioneer study through an examination of the sources and evaluation of environmental issues.

The environmental impact of industrial activity was discernible in the present-day territory of Slovakia as early as the High Middle Ages, primarily in connection with intensive mining and metallurgical activity. As was the case in other countries, this impact grew substantially more serious with the rise of industrialism at the end of the nineteenth century. It became evident mainly in the mining regions of central and eastern Slovakia with centers of ore processing via chemical-metallurgical processes. This period saw the foundation of several industries with major ecological impacts, including the cellulose-paper industry in Central Slovakia with immediate waste discharge into rivers; the production of cement and lime with intensive open-pit mining; the chemical manufacturing of explosives, sulfuric acid and superphosphates in the Dynamit Nobel plant in Bratislava and elsewhere; and the Apollo oil refinery in Bratislava. In the southern parts of central and eastern Slovakia, a network of plants specializing in the extraction and processing of magnesite was created. More effective measures intended to mitigate the impact of these industries were implemented in the interwar period and at the beginning of the 1940s. The measures, however, were unsystematic and the results were very modest.[4]

A historic turn in the development of industry structure in Slovakia was brought about by the first Five-Year Plan and particularly the new orientation toward the military–industrial complex, including heavy engineering, metallurgy, chemical engineering, and other materially and energetically demanding branches of industry. This new emphasis affected the economy of the whole Czechoslovak Republic (CSR), but it had an even greater impact on the Slovak economy, whose industrial structure had only just begun to form at the time. The reorientation of the economy in the 1950s therefore gave the Slovak economy a certain character, which essentially persisted until 1989. The government concentrated in Slovakia the manufacturing of metallurgical and ferrous semi-finished products, various chemical semi-finished products, oil products, plastics, and the production of heavy weapons. Great demands in terms of material and energy inputs necessitated the construction of large thermal power plants and a system of hydroelectric power plants with valley dams, leading to far-reaching effects on the countryside. The environmental consequences of this economic development were only marginally studied in the mid-1950s.[5]

Typical examples of the impact of large socialist construction projects on the landscape include the construction of an aluminum plant in the town of Žiar nad Hronom and the petroleum refinery Slovnaft complex in Bratislava. The production of fuels and aluminum, especially in the second half of the twentieth century, was strategically essential for all the world's most advanced countries. Construction for aluminum production with large energy sources was already being prepared by the German-controlled independent Slovak Republic during the war years. Given the economic and political conditions of Stalinist industrialization, the construction of large central plants took on a completely new dimension with far-reaching effects. This was also true for communist Czechoslovakia. The Soviet Union became involved in the planning of large-scale industrial plants, or initiated their construction in their own interests; the capacity of the plants was generally oversized because they fulfilled a strategic role for the whole Soviet bloc and its military–industrial complex. The planning and construction of industrial plants proceeded hurriedly and chaotically. Technology was already behind that of the advanced Western countries in the design phase; the location of the plants often did not respect the conditions of the natural environment and human habitation; and the negative environmental impacts on nature and man were taken into account only where absolutely necessary or were not taken into account at all.

The genesis of the aluminum production plant in Žiar nad Hronom (formerly known as Svätý Kríž nad Hronom) had all the basic attributes of the development of large projects of socialist industrialization during the Stalinist period in central Europe. Czechoslovakia needed aluminum for both military and civilian industry, but the main raw material, bauxite, was not available for production, and the extremely high energy demand of aluminum production was also a problem. The hurried construction of heavy industry in Czechoslovakia in the early 1950s created an energy crisis. The government regime faced a dilemma about whether, or under what conditions, to start producing aluminum or importing it from abroad. However, due to the economic embargo of Western countries, there was not a wide choice of suppliers. The Czechoslovak government's dilemma was solved by Soviet leaders in cooperation with the communist regime in Hungary under the leadership of one of the most radical Stalinists, Mátyás Rákosi. Hungary had large bauxite reserves but lacked the suitable technology, industrial capacity, and energy resources needed to process it. Probably for the reasons mentioned, the Soviet authorities decided that a substantial increase in aluminum production for the needs of the countries controlled, especially for defense purposes, would be provided by Czechoslovakia as industrially and technologically the most developed country of the socialist camp. The choice could have been influenced by the fact that the core of the military–industrial complex of the communist states of central Europe was being built in Czechoslovakia. Within Czechoslovakia, weapons, especially heavy arms, production was concentrated mainly in Slovakia, as it was a hinterland to an expected military attack from the West. That is why the new aluminum plant was to be located in Slovakia in its central part, at a sufficient distance from the border.[6]

The proposal (or rather the directive) to build an aluminum factory in Czechoslovakia was officially delivered by A. I. Mikoyan, the Soviet national economist and close associate of J. V. Stalin, in November 1950 at the meeting of the Soviet bloc economic council of the Council of Mutual Economic Assistance (COMECON). The new complex was to be designed and built on the basis of Czechoslovak–Hungarian cooperation with hidden control by the Soviet Union, which was to act as coordinator and also arbiter of possible disputes. The construction of the plant was pushed forward by Rákosi, the most active supporter of Stalin's methods among the representatives of Soviet satellites in central Europe. He also advocated the interconnection of the railway system of Czechoslovakia with the Soviet Union's wide-gage railways and the construction of a mammoth goods transshipment yard at the Czechoslovak–Soviet border in Čierna nad Tisou.[7]

The highest functionary of the Czechoslovak communist regime and the president of the Czechoslovak Republic, Klement Gottwald, competed with M. Rákosi for the favor of the Soviet leadership and Stalin personally. He could not afford to reject Soviet economic requests, especially when they were of a military-strategic nature. This would jeopardize its position as well as the position of Czechoslovakia in the Soviet bloc. The political leadership of Czechoslovakia was probably familiar with the plans for aluminum production or anticipated them, because the Research Institute of Metals in Prague had already prepared estimates of the financial costs and raw materials inputs of the future plant in mid-1950. In January 1951, the economic representatives of Hungary and the Czechoslovak Republic, under the leadership of the planning offices of the two countries, Zoltán Vas and Jaromír Dolanský, began negotiations on the conditions for the construction of the plant. Due to the economic and political relations of the period and the dominant ideology, the negotiations did not focus on establishing capital or technological cooperation between two states, but rather on arranging economic aid from Hungary for the aluminum industry in Czechoslovakia. Hungary had to supply project documentation, patents, technology facilities, and experts to teach Czechoslovak technicians about bauxite processing. At the same time it was supposed to supply bauxite and aluminum oxide. Czechoslovakia would in return supply equipment for power and heating (which was still at world-class level), high quality metallurgical coke, and strategic goods such as ball bearings.[8]

However, Hungary only had the technology to process the most valuable bauxite with the grade expressed in "modules" of ten to fourteen, while it was willing to export to Czechoslovakia only less valuable bauxite with a grade of only six to eight modules. The necessary technology for the processing of low-quality raw materials would be developed by the Czechoslovak Republic with the help of Soviet specialists, or by purchasing patents from Western countries in cooperation with Hungary. In reality, the technological assistance of Hungary could only be used partially. The role of Czechoslovakia was to spend a great deal on the development of new technology and then produce aluminum from low-value raw materials with disproportionately high energy inputs and with considerably more serious environmental impacts than would be the case when processing higher quality bauxite. Engineers and technicians had minimal time for preparation of the project, for development of relevant technology, and for construction. In April 1951, the Central Committee of the Communist Party and the Government of the Czechoslovak Republic made the commitment to build the plant and in November 1951 construction began. There were several ideas for the location of the plant. Based on the advice of the Soviet experts, they eventually chose a site in the Hron valley in central Slovakia in the town of Svätý Kríž nad Hronom, which served as the summer residence of the Bishop of Banská Bystrica. Due to strict standards of secrecy at the time, we still do not know the details of the construction preparation. Most likely, however, the influence on nature, the country, and the population was not considered. The huge ecological costs of construction and the necessity of importing technologies from advanced Western countries already excluded such considerations. This was evidenced by the fact that in the first stages of construction, emission separators, water purifiers and other elements to reduce ecological effects were all missing.[9]

The construction of the aluminum plant progressed in several phases. In the first phase, which began and ended in 1954, several electrolysis plants to process aluminum oxide from Hungary were built. In the second phase, lasting until 1957, additional plants were added, making it possible to process low-module bauxite and produce aluminum oxide for final aluminum production. However, the import of bauxite and temporarily also aluminum oxide required the expansion of the capacities of bauxite mining and processing plants in Hungary. Czechoslovakia had to provide Hungary with the necessary energy facilities for the expansion of capacities and a large part of the electricity supplies. As Hungary lacked the financing to buy equipment, Czechoslovakia had to provide a loan for the purpose, which was then repaid with Hungarian bauxite and aluminum oxide. The main purpose of building the plant, that is, aluminum production, remained secret until the start of the first operations, even from the ordinary employees. That is why individual parts of the works were given cover names such as "Youth Building," "Engels I," and so on. The new plant was named Kovohuty Hron and after 1953, Závody Slovenského národného povstania (The Slovak National Uprising Works). In the initial period after the start of production, experts found that emissions from plant operations contained hazardous substances and gases, such as carbon monoxide, sulfur dioxide and, in particular, fluorine, all of which were capable of seriously damaging the environment.[10]

Emissions affected the land up to 15 kilometers away. As a result, the aluminum industry became the first case that forced the Czechoslovakian government to address environmental issues related to large industrial complexes. The effects of aluminum production began to be addressed by the authorities of the Communist Party, state authorities, and scientific institutions. The experience of other countries has shown that the ecological impacts of construction are best addressed in advance, with plants sited in open country, far from human settlements and on less valuable land, with adequate means for the dispersion of emission. However, the Kovohuty Hron aluminum plant was placed in a valley between two mountain ranges near Žiar nad Hronom, on valuable agricultural land.[11]

To reduce the leakage of hazardous substances from the plant into the air and nearby water courses, it was necessary to use exhaust and gas neutralizing equipment, as well as a treatment plant and an underground wall for the protection of groundwater. In the mid-1950s, the authorities decided to install exhaust systems, but they did not enter operation until 1966. Likewise, the water treatment plant and the underground wall at the plant did not operate until the 1980s. These partial environmental measures lagged far behind the development of aluminum production, which rose from 2.8 thousand tons to 70 thousand tons between 1953 and 1975, making the plant the largest of its kind in central Europe. The number of inhabitants of Žiar nad Hronom increased tenfold from the end of the 1940s to the 1980s, from 1.5 thousand to 15 thousand. As the aluminum works operated without effective environmental protection until the mid-1960s, the environment and the population in the wider neighborhood were severely affected by dangerous substances. In one monitored period, electrolysis at the plant produced 2400 kg of fluorine per day. The small village of Horné Opatovce, which was only 200 meters away from the works, was affected so much that the authorities decided to evacuate all of its 1,096 inhabitants.[12]

Emissions from the aluminum plant gradually affected 3,630 hectares of agricultural land and seriously damaged the nearby forested highlands of the Štiavnické vrchy and Kremnické vrchy, especially the vegetation of a special kind of Slovak fir, considered to be the highest quality and the most beautiful kind of fir in Europe. Slovak fir trees were also threatened throughout the entire territory of Slovakia. Its share in the total area of Slovak forests fell from 11 percent to 5 percent from the 1950s to the 1980s. In addition to the plant, the mountain of production waste, which in the 1980s reached an area of forty hectares and a height of forty-five meters, gradually grew. It contained about 8 million tons of sludge and about 1 million cubic meters of alkaline water. The flow of contaminated water into the subsoil contaminated the flow of the Hron River.[13]

On arable land it was no longer possible to grow cereals, maize, or other more demanding crops, but instead only potatoes and flax. The milk production of dairy cows in the region fell to about a third of the national average, and livestock feed had to be imported from other regions. At the instigation of experts, a state farm institution was founded on the damaged soil around the plant, where they investigated the effects of emissions on crops and looked for ways to recultivate the land. However, it was determined that recultivation would require extraordinarily high costs that the state could not afford. The employees of the plant and the inhabitants of Žiar nad Hronom began to suffer from fluorosis: according to some unverified data, up to about 90 percent of workers from the aluminum works contracted cancer over time.[14]

Another ecological artifact from the period of socialist industrialization in Slovakia is the mammoth Apollo oil refinery built on the outskirts of the capital of Bratislava, nowadays inside the boundaries of the city. As already mentioned, the refinery was first established in Bratislava at the end of the nineteenth century. It served the needs of the Kingdom of Hungary (which at that time included the area of modern Slovakia) and imported raw materials from the north from the region of Halič, which at the time belonged to the Habsburg monarchy. The plant was built in the industrial zone of the city near the river port on the Danube. In terms of the current area of Bratislava, the refinery is located in the wider center of the city, where over fifty years of operation without protective measures deposited a large volume of oil and its products into the subsoil. In June 1944, however, the refinery fell victim to an allied air raid, in which it was almost completely destroyed.[15]

After the end of World War II, the refinery was partially rebuilt and production restored. However, the government had already decided to move it. First they planned to build a new works further from the border in central Slovakia near Trenčín. At the end of the 1940s, the plan was altered, and the refinery was once again set for construction in Bratislava, but further from the city center, in Vlčie Hrdlo. The construction of the plant was carried out during the socialist industrial drive of the 1950s, along with the Žiar nad Hronom aluminum works, and began in August 1957. Initially, the processing capacity was 200 thousand tons of oil per year. However, the completion of the new Družba (Friendship) pipeline from the Soviet Union to Slovakia in 1962 enabled a gradual increase in the planned processing capacity by several hundred percent.[16]

The project plans for the new refinery named Slovnaft did not even consider protection against oil leakage or the construction of a water treatment plant, although the risk of oil spills was greater because the plant was built on the permeable gravel floodplains of the Danube, where large supplies of good drinking water are located. The 4 km^2 refinery was located on the edge of a very large river island, known as Žitný ostrov (Rye Island), created by the Danube River and its tributary the Little Danube, under which there is an underground lake fed by Slovak rivers, classified by the UN as among the world's major drinking water sources. Oil spills were recorded in the 1950s, but government authorities did not pay attention. The problem began to be dealt with only in 1961, when refinery petroleum polluted one of the major drinking water sources in the nearby village of Podunajské Biskupice, which is now part of Bratislava. However, until an effective solution was found, oil leaked further into the underground waters and created a huge slick on the underground lake. A special hydraulic curtain system was developed to remove it. It was formed from a ring of hundreds of wells drilled around the refinery, which pumped water and oil from the surface of the underground lake. The system began operating in 1973 and by the mid-1980s it pumped out a large amount, about 75,000 thousand cubic meters of oil. According to official data, this underground spot had shrunk to about half the original area. Another task was to build a wastewater treatment plant for the refinery. However, this was not completed until 1986.[17]

Specific data about the state of the environment with the rise of socialist industrialization in the 1950s are available only indirectly, mostly from the following period, because everything related to the development of the economy was concealed at the time. Among the biggest polluters during the period of socialist industrialization were the Kovohuty (non-ferrous metal works) in Istebné and Široké in North Slovakia, the nickel smelting plant in Sereď in West Slovakia, the East Slovakian Ironworks in Košice, and the iron ore mines in Rudňany in Northeast Slovakia. The first direct pollution datasets are available from 1961. According to these data the aluminum plant in Žiar nad Hronom produced as much as 8,650 metric tons of fluorine per year; the Nováky, Žilina, and Handlová chemical plants in Middle Slovakia produced 7,680 tons of chlorine, sulfur dioxide and other emissions; the chemical plant of J. Dimitrov (previously Dynamit Nobel) in Bratislava produced 8,813 tons of hazardous gases; the Jelšava, Košice and Hačava magnesite plants in East Slovakia emitted 30,300 tons of exhaust emissions; and the thermal power plants Nováky, Zemianske Kostoľany, and Handlová plants produced 122,950 tons of flue-ash and arsenic compounds.[18]

The overall data for Slovakia were not published until 1972. These sources mentioned the overall production of emissions equaling 1.285 million tons of emissions, of which 0.597 million tons were solid and 0.688 million tons were gas exhausts. The biggest producers (33 percent) of the solid emissions were thermal power plants and boiler houses of industrial enterprises, while 27 percent was related to energy production and 10 percent to ferrous and non-ferrous metallurgy. The biggest producers of sulfur dioxide were again power plants and boiler houses of plants, a 59 percent share. Harmful pollutants affected 12 percent of the population, 120,000 hectares of arable land and 37,000 hectares of forested land. Among the most affected regions were Bratislava, the Nitra river valley (Nováky, Handlová), the Váh river valley (cement and cellulose plants), the Hron river valley (Žiar nad Hronom), the Gemer region in Southeast Slovakia (magnesite), the Spiš region in Northeast Slovakia (ore processing), and Košice; in all, the affected areas represented about a third of Slovakia.[19]

The ecological consciousness of the Slovak public, and also of those in governmental structures, emerged during the 1960s. During the Third Five-Year Plan (1961–1965), the first ambitious program for the removal of pollutants by means of professional committees of environmental protection was adopted. These committees were supposed to cooperate with the plant managers. The fulfillment of the program, however, fell behind targets and the program eventually disintegrated along with the entire Third Five-Year plan. Environmental protection measures to mitigate the impact of manufacturing activity reached the legislative level through Act no. 35 in 1967, which introduced penalties and fees for air pollution. The first permanent authority for the function of environmental character was established: the State Technical Inspection of Air Protection. Similar tasks were also fulfilled by the Hydrometeorological Institute and the state hygienic services.[20] The federalization of Czechoslovakia in 1969 brought the next changes.

The federation consisted of Czech and Slovak republics with separate, national ministries responsible for the protection of the environment. In the following years, the Slovak governmental bodies started to develop systematic environmental policies. However, this process was hampered by the strategic economic concerns of the state and the problems of the status of the Slovak republic in the Czechoslovak federation. A systematic approach to environmental protection started during the 1970s under the influence of the worldwide environmentalist movement, a branch of which appeared in Slovakia. The Slovak government, as was also the case with the Czech government, featured an Environmental Council. The years from 1972 to 1974 saw the elaboration of a wider program of environmental protection and development, approved under the title *Summarized Concept of Environmental Care in Slovak Socialist Republic in Relation to the Sixth Five-Year Plan (1976–1980)*. The preparation of this document also included the analyses of the aforementioned data and conclusions about the impact of industrialism on nature, the countryside, and the quality of life.[21] The *Summarized Concept* was in line with the environmental policy strategy of the central government in Prague and was a solid part of it. It enabled the elaboration of the main environmental objectives set by central state authorities in the 1970s on the specific conditions of Slovakia. Special environmental programs have also emerged for the needs of Czech countries.

The Slovak long-term environmental protection plans officially referred to the resolutions passed by the highest Soviet party authorities, similar to the complex nature of the socialist industrialization of Slovakia and the Czechoslovak Republic, harmonized with the joint plans of the countries included in the COMECON. Specifically, the *Summarized Concept* aligned with the conclusions of the Twenty-Fourth Congress of the Soviet Communist Party and the Fourteenth Meeting of the Communist Party of Czechoslovakia in 1971, as well as conclusions and materials from the numerous sessions of the Slovak Government. The *Concept* began with the obligatory cited speech by Leonid Brezhnev at the Twenty-Fourth Party Congress about environmental protection. The speech was mostly rhetoric, but it confirmed the fact that the top Soviet representatives acknowledged the seriousness of the situation and the inevitability of adopting systematic environmental measures. Socialist countries initiated the convention of an international symposium in 1967 on environmental protection. The symposium was held in 1971 in Prague with the participation of a few Western and developing countries within the activities of the UN European Economic Board. The Soviet Union and its satellite countries established the Environmental Board, a special joint body at COMECON.[22]

Given the conditions in Czechoslovakia, the Twenty-Fourth Party Congress signaled a departure from the liberalizing initiatives begun at the end of the 1960s, known as the Prague Spring. However, the Party Congress also inaugurated a new era of higher environmental protection. For example, the aforementioned Environmental Boards within the Slovak and

Czech Governments were established on the basis of its resolutions. Environmental issues were further elaborated to meet the specific conditions of Slovakia. The Slovak National Council Board for Environmental Plan and Budget was established in the Slovak Parliament. A special fund was created concurrently, aimed at funding the measures to be adopted in the stated area. The provision of funds gave hope that the environmental plans could be fulfilled. The Slovak Government and the Slovak National Council empowered the ministries and regional administration bodies (the regional national committees, or KNVs) to address environmental issues according to particular areas of the subject matter and the situation in particular regions. Knowledge acquired from subordinate bodies was transformed by the Slovak Government to particular resolutions passed and materials elaborated on the key problems during its meetings held during the period of 1971 to 1973. In particular, resolutions were passed to ensure the cleanliness of surface waters, the protection of the Žitný ostrov aquifer, to improve the sanitary quality of workplaces, and to improve the task of nature protection. Based on these resolutions and further analyses prepared by the ministries, the Slovak Government, through the Ministry of Construction and Technology, produced the *Summarized Concept* in January 1974. The government warned in advance that the plan dealt only with the most major and urgent problems of nature protection, despite the name, thereby admitting that the document was not capable of addressing the entire spectrum of contemporary environmental issues.[23]

The *Summarized Concept* comprehensive plan was primarily focused on environmental conditions related to air quality, surface and underground waters, forest and soil protection, noise pollution, solid waste disposal, and overpopulation, as well as the improvement of working conditions and recreational facilities. A state authority was assigned to each area, usually a ministry. In addition, certain plans were given deadlines, as well as suggested solutions and proposed methods for assessment of success or failure. One of the main intentions of the *Summarized Concept* was to highlight unacceptable forms of economic and industrial development, such as the energy sector and other industrial sectors combusting solid fossil fuels. It is important to note that the *Summarized Concept* was strictly confidential and prepared in two variants: one shorter and of a more general nature, containing more favorable data, and the other more detailed, containing more accurate and more alarming analyses.[24]

The *Summarized Concept* noted the regions of Slovakia that were most affected by environmental devastation, and where the main ecological measures should be directed. They were the regions of Bratislava, the basins of the rivers Nitra, Váh and Hron (Ponitrie, Považie, Pohronie), the Spiš region in Northeast Slovakia, and the area in the east of Slovakia between the city of Košice and the town of Jelšava. In the 1970s, 60 percent of the Slovak population lived in these regions. In the area of air protection, the *Summarized Concept* focused mainly on the problem of collecting waste gas containing solid pollutants, especially fly ash and dust generated by thermal power plants, heat plants, cement plants, magnesite, and other production plants, and also waste gas containing sulfur dioxide, arsenic, zinc, fluorine, and other substances released from chemical plants.[25]

According to the shorter summary versions of the *Summarized Concept*, the capture of solid pollutant emissions using electrostatic separators and other methods had been mostly implemented (80 to 95 percent of the time in most cases), especially in case of fly ash from the boiler rooms of large power plants and heating plants, and even 99 percent in some locations. The capture of fly ash released by smaller boiler facilities and the separation of dust from cement, magnesite, and metallurgy plants was less successful, despite the installation of suitable equipment. The aluminum plant at Žiar na Hronom represented an exception, with installed fluorine waste gas separators, although with 50 percent efficiency instead of the declared 75 percent. The waste gas exhausted was separated successfully in approximately 80 percent of industrial plants, as well as from industrial fertilizer production and the recovery of petrol and carbon disulfide vapors.[26]

The environmental outlook for the Sixth Five-Year Plan assumed the installation of improved high-performance separators, electrostatic filters, and new types of de-dusting equipment intended to catch waste gas and other solid pollutants in the atmosphere. Despite separation intensification, the amount of particulate matter emissions was projected to remain consistent throughout the 1970s, because the economic plans foresaw increased expansion of thermal power engineering capacities and other low-quality coal-based operations, at least until 1980. In fact, the capacities increased beyond 1980. Lignite mined from Czech and Slovak mining regions contained three to four times higher ash content and higher sulfur content, but only approximately half of the energy value compared to coal mined in other countries. The government proposed a solution: to accelerate the construction of nuclear power plants at Trnava and Mochovce, although these plans presented a highly controversial solution with unforeseen and far-reaching effects, especially considering their reliance on obsolete Soviet technology. Rural gasification plans represented one possible alternative path, especially in the area of boiler power plants, while increasing gas imports from the Soviet Union to a level of at least 4 billion m^3 a year also was stated as feasible.[27]

While the government planned to maintain the status of the 1970s in the solid pollutant emissions segment, the amount of released gas pollutants was set to increase by 50 percent by 1980, as a result of the Sixth Five-Year Plan and the expected expansion of heating systems, chemical, metallurgical, and other industrial sectors, and an expected boom in motorized transport. In order to reduce sulfur dioxide emissions, the installation of desulfurizing equipment in the oil refinery Slovnaft Bratislava was planned, allowing for the reduction of sulfur content in oil from approximately 1.7 percent to 0.5 percent. Taking into account the planned increase in the processing capacity of the Slovnaft refinery to 13 million tons of oil per a year by the year 1980, a second stage of desulfurizing equipment had to be considered in advance. In addition to expensive ecological projects, the simple observance of existing standards, disregarded by most companies on a regular basis, would have been sufficient to produce a major elimination of pollutants. Therefore, systematic inspections of industrial plants were planned. The assessment of pollution abatement measures required advanced monitoring devices that could only be imported from Western countries, but the state did not have enough funds to purchase such devices.[28]

In the terms of water protection, the huge underground water reserves in Slovakia were determined to contain a total volume of 50 m^3/s, but these were distributed unevenly and as a result 60 percent of Slovak territory reported a shortfall of underground water sources—the best source of drinking water for households. It was deemed necessary to connect the deficit regions with the largest underground water reservoir in Slovakia, near Žitný ostrov. This would in turn require the strict protection of the underground lake from pollution by the Slovnaft refinery. Along with drinking water, Slovakia needed a huge amount of water for industrial, agricultural, and recreational uses. The *Summarized Concept* anticipated an increased water demand from 1.380 million m^3 in 1970 to 4.07 million m^3 in 1990. Such water consumption could be satisfied only through the use of surface water reserves capable of providing approximately 3.4 million m^3 of water per a year. However, 1,400 km of Slovak rivers, with a total length referring of approximately 5,100 km were strongly contaminated, corresponding to the highest contamination category, level IV. The rivers Váh, Nitra and Hron were those most affected but other Slovak rivers were also seriously contaminated.[29]

The government wanted to establish six regional water management systems that would be mutually interconnected through the newly built water conduits structure, in order to ensure drinking water supplies from underground water sources. They would deliver drinking water from the reservoir underneath Žitný ostrov to those parts of central and eastern Slovakia that lacked underground water reserves. Further locations in various regions of Slovakia would be gradually connected to these water conduits in the more distant future. As a result, the number of towns and villages connected to the water conduit would increase from 35 percent in 1974 to 85 percent in 2000. However, the fulfillment of this objective was planned to begin after the year 2000. The use of surface water for economic and recreational purposes required the treatment of contaminated water in water courses assigned to contamination category IV. The volume of wastewater sluiced into the Slovak rivers per year was expected to increase from 1.137 million m^3 in 1970 to 2 million m^3 in 1990. The resolution of this situation depended upon intense wastewater purification so that the share of contaminated water in the total volume of water sluiced in the rivers would be reduced from 58 percent in 1970 to 26 percent in 1990. To fulfill these plans, the construction of approximately 500 water purification plants equipped with multi-level water purification technology were installed, with the purification level gradually increasing to 90 percent. The construction of water purification plants should have been associated with the adoption of further measures, including sludge disposal and sewage systems refurbishment, but these were not included.[30]

The protection of agricultural and forested lands was a very complicated task since Slovakia had only 2.623 million hectares of undeveloped land available in the mid-1970s, but nevertheless the second largest forest area available in Europe after Sweden, 1.852 million hectares. However, given the conditions of socialist industrialization, the amount of agricultural land had been quickly reduced and future forecasts were not optimistic. Slovakia lost 133,000 hectares of agricultural undeveloped lands during the period of 1960 to 1972 and an additional 173,000 hectares of agricultural lands were expected to be lost to development by 1990. This reduction was attributed to the expansion of built-up areas, increasing investment projects, mining, and the expansion stockpiles of exhausted land. Most affected were, as in the case of gaseous emissions, the regions of Bratislava, the Nitra river valley (Nováky, Handlová), the Váh river valley (Žilina, Ružomberok), the Hron river valley (Žiar nad Hronom, Banská Bystrica), the Gemer region, the Spiš region and Košice. Further forecasts foresaw the expansion of water surfaces and forest planting on the exhausted lands.[31]

The agricultural areas harmed by solid and gaseous pollutant emissions, as well as other adverse ecological effects, numbered at approximately 120,000 hectares. Industrialized agriculture had increased the amount of exhausted land subsequent to the intense application of chemical fertilizers and chemical herbicides. During the period from 1937 to 1973, the application of chemical fertilizers had increased from 5 kg to 193 kg per hectare, and an additional increase to 346 kg/hectare was planned for 1990. Harmful substances were widely used as herbicides and pesticides in the postwar period, for example chlorinated hydrocarbons such as DDT. Later, these substances were restricted, but the absorbed traces of chemical herbicides and pesticides caused considerable damage to soil, surface and underground water, micro flora, and last but not least also humans.[32]

The huge increase in animal production that boomed after agricultural collectivization and the creation of agricultural associations in Slovakia also had serious consequences. For example, the breeding of 15,000 hogs annually generated 27,300 m^3 of liquid and solid waste per year, and a thousand head of beef cattle generated 3,400 hectoliters of liquid manure. Hundreds of hectares of land were used for liquid excrement disposal but soil contamination and penetration of pollutants into the aquifers could not be avoided. The construction of thirty-six new hog breeding farms was planned for the Sixth Five-Year Plan, each with 500 to 1,500 animals, and ten farms would have at least 5000 hogs. Stockpiles and de-sludging yards represented another dangerous phenomenon. Approximately 700 officially registered stockpiles and de-sludging yards with a total area of 1,100 hectares existed in Slovakia in the mid-1970s and the area was planned to increase to 1,800 hectares by 1980.[33]

The government enacted stricter terms of land allocation for various purposes and the reclamation of damaged lands with the aim of fighting against the rapid reduction of agricultural land area. These measures came at a high financial cost. One of the ways identified was the intensification of agricultural production, including the application of new, safer forms of chemical fertilizers, and changing the structure of agricultural companies. However, agricultural intensification proceeded as before, with mostly traditional methods dominating.[34]

A positive trend of gradually expanding forest land was reported by forestry agencies. A total of 58,000 hectares of forest were added in Slovakia during the period from 1960 to 1974, with an estimated 90,000 additional hectares of forest land by 1990. Based on the national economic plans, the new forest lands were to serve mainly recreational and other cultural and social purposes. The expansion of traditional forest management was planned in order to eliminate soil efflorescence and dry-out which, left unchecked, would result in soil erosion and landslides, and also to fulfill other water management, climatic and other tasks. Waste gas containing fly ash, dust, sulfur dioxide, arsenic, fluorine, and other substances, continued to affect forested areas, and large areas of forest were exposed to acid rains. In the mid-1970s, the area of forest damaged by air pollution was estimated in approximately 19,000 hectares in the area of Žiar nad Hronom, the Upper Nitra River region with large thermal power plants, and in eastern Slovakia around the magnesite industry centers of Jelšava and Hačava.[35]

The fulfillment of the tasks set forth in the *Summarized Concept* was reviewed on an annual basis during the Sixth Five-Year Plan, with a final assessment for the whole five-year period in 1980. Similar summary concepts of environmental protection with forecasts through the year 2000 were prepared during the period of the Seventh Five-Year Plan (1981–1985) and the unfinished Eighth Five-Year Plan (1986–1990). These documents, as well as the Environmental Protection Tasks Fulfillment Report during the Sixth Five-Year Plan, contained much less specific information than the *Summarized Concept*, and were mainly of a general nature. Taking into account the contents of these documents, we could state that the environmental protection plans from the mid-1970s were only partly fulfilled. Compared with the data from the *Summarized Concept*, the further aggravation of environment in the area of atmosphere, surface and underground waters, and soil and forest protection was slowed or even eliminated in certain areas. Based on UN European Economic Board data collected in 1982, Slovakia ranked third in Europe in air pollution per capita.[36]

Epilogue

The further economic development of Slovakia and the Czechoslovak Federation after the mid-1970s served to preserve only an ecologically disadvantageous structure. There was a partial shift from extensive forms of economic development, but the radical reduction of energy and material inputs and the rise of new technologies, characteristic of Western countries of the era, as well as cybernetization, informatics, and automatization, came to CSR and Slovakia only with a delay and with a decreased effect on production intensification. Therefore, the growth of emissions and other forms of ecological impact could be slowed, but not stopped or reduced.

Notes

1 Mikuláš Huba (ed.), *Slovensko: Slovenské lesy-otázka existenčná (Slovakia: Slovak Forests-Question of Existence)* (Bratislava: Slovenský zväz ochrancov prírody a krajiny, 1990); Mikuláš Huba (ed.), *Ponovembrové Slovensko vo vzťahu k životnému prostrediu a trvalo udržateľnému životu (Slovakia after November in the Relations to Environment and to Continual Sustainable Life)* (Bratislava: EuroUNiPress, 1994); Mikuláš Huba and Vladimír Ira, *Stratégia trvalo udržateľného rozvoja vo vybraných regiónoch (Strategy of Continual Sustainable Development in Selected Regions)* (Bratislava: Spoločnosť pre trvalo udržateľný rozvoj, 2000).

2 Ján Mlynárik, *Ekológia po Slovensky: Otázky životného prostredia na Slovensku 1948–1989 (Ecology According to Slovakia: Questions of Environment in Slovakia 1948–1989)* (Prague: Danubius, 1994).

3 Michal Barnovský, "Industrializácia Slovenska a životné prostredie v období komunistického režimu (Industrialization of Slovakia and Environment in the Communist Era)," *Acta Oeconomia Pragensia*, 2007, 15, no. 7: 55–71.

4 Ľudovít Hallon, "Charakter environmentálneho myslenia a opatrení v priemysle Slovenska 1918–1938: Celkový stav a názorné príklady (The Nature of Environmental Thinking and Measures in Slovak Industry from 1918 to 1938: General Condition and illustrative Examples)," *Forum historiae: odborný internetový časopis pre históriu a príbuzné spoločenské vedy (Forum historiae: professional journal for history and related socially sciences)*, 2017, 11, no. 1: 112–127.

5 See, for example, Michal Barnovský, "Problematika industrializácie Slovenska v rokoch 1945–1950 (Problematic of Industrialization of Slovakia In The Years 1945–1950)," *Historický časopis (Historical Journal)*, 1968, 51, no. 2: 169–190; Miroslav Londák, *Otázky industrializácie Slovenska 1945–1960 (Questions of Industrialization of Slovakia 1945–1960)* (Bratislava: Veda, 1999); Miroslav Londák, *Ekonomické reformy v Československu v 50. a 60. rokoch 20. storočia a slovenská ekonomika (Economic Reforms in Czechoslovakia in the 1950s and 1960s Years of 20th Century and Slovak Economy)* (Bratislava: Historický ústav; Vydavateľstvo Typoset, 2012).

6 "Korešpondencia ku vzniku hlinikárne v Žiari nad Hronom (Correspondence about generation of aluminum plant in Žiar nad Hronom)," in Národní archiv Praha (National Archive Prague), Fund 100/24, Band 96. Elaborated in Ján Mlynárik, *Ekológia po Slovensky: Otázky životného prostredia na Slovensku 1948–1989 (Ecology According to Slovakia: Questions of Environment in Slovakia 1948–1989)* (Prague: Danubius, 1994), 150–157.

7 Ján Mlynárik (1994), 150–157.

8 Ján Mlynárik (1994), 150–157.

9 Ján Mlynárik (1994), 150–157; Peter Ratkoš (ed.), *Dějiny Žiaru nad Hronom. (History of Žiar nad Hronom)* (Martin: Osveta, 1978), 218.
10 Ján Mlynárik (1994), 150–157; Peter Ratkoš (1978), 219–227.
11 Michal Barnovský, "Industrializácia Slovenska a životné prostredie v období komunistického režimu (Industrialization of Slovakia and Environment in the Communist Era)," *Acta Oeconomia Pragensia*, 2007, 15, no. 7: 59–60.
12 Ján Mlynárik, Prečo umiera slovenská jedľa? (For What Reason is Slovak Fir Dying?)," in Ján Mlynárik (1994), 12–14; Michal Barnovský (2007), 60–61.
13 Ján Mlynárik (1994), 12–14; Žiarska dolina—SOS, SOS, SOS ... (Žiar Valley—SOS, SOS, SOS...) in Ján Mlynárik (1994), 24–26; Michal Barnovský (2007), 60–61.
14 Magda Rybárová, "Žiarska dolina volá o pomoc (Žiar Valley Calls For Help)," *Roľnícke noviny (Farmers Journal)*, January 16, 1984; Ján Mlynárik (1994), 24–26; Michal Barnovský (2007), 60–61.
15 Ľudovít Hallon and Miroslav Londák, *Chemický priemysel v zrkadle dejín Slovenska. (Chemical Industry in The Mirror of The Slovak History)* (Bratislava: Nakladateľstvo STU v. Bratislave, 2011) 21–22, 29–30 and 40.
16 Ľudovít Hallon and Miroslav Londák (2011), 46–54 and 58; Michal Barnovský (2007), 61.
17 Kauza Slovnaft (Cause of Slovnaft) Ján Mlynárik (1994), 47–48; Ľudovít Hallon and Miroslav Londák (2011), 46–54 and 58; Michal Barnovský (2007), 61.
18 Session of Chairmanship of Central Committee of Communist Party of Slovakia, December 21, 1962. Report about pollution of air in selected industry plants in Slovakia in the year 1961, in Slovak National Archive (SNA), fund Central Committee of Communist Party of Slovakia, box 1, 096, fascicle 475/15.
19 Session of Chairmanship of Central Committee of Communist Party of Slovakia, July 9, 1974. Summarized concept of environmental care in Slovak Socialist Republic in relation to the Sixth Five-Year Plan (1976–1980), in SNA, fund Central Committee of Communist Party of Slovakia, box 1 390, fascicle 1 334/19.
20 Michal Barnovský (2007), 61–62.
21 Session of Chairmanship of Central Committee of Communist Party of Slovakia July 9, 1974. Summarized concept of environmental care in Slovak Socialist Republic (SSR) in relation to the Sixth Five-Year Plan (1976–1980), 69–71, in SNA, fund Central Committee of Communist Party of Slovakia, box 1 390, fascicle 1 334/19.
22 Summarized concept of environmental care in SSR in relation to the Sixth Five-Year Plan (1976–1980), 68–69, in SNA, fund Central Committee of Communist Party of Slovakia, box 1 390, fascicle 1 334/19.
23 Summarized concept of environmental care in SSR in relation to the Sixth Five-Year Plan (1976–1980), 7–9, 30–31, in SNA, fund Central Committee of Communist Party of Slovakia, box 1 390, fascicle 1 334/19.
24 Summarized concept of environmental care in SSR in relation to the Sixth Five-Year Plan (1976–1980), 9–10, 71–73, in SNA, fund Central Committee of Communist Party of Slovakia, box 1 390, fascicle 1 334/19.
25 Summarized concept of environmental care in SSR in relation to the Sixth Five-Year Plan (1976–1980), 10–11, 72, in SNA, fund Central Committee of Communist Party of Slovakia, box 1 390, fascicle 1 334/19.
26 Summarized concept of environmental care in SSR in relation to the Sixth Five-Year Plan (1976–1980), 11–12, 74–83, in SNA, fund Central Committee of Communist Party of Slovakia, box 1 390, fascicle 1 334/19.
27 Summarized concept of environmental care in SSR in relation to the Sixth Five-Year Plan (1976–1980), 11–12, 83–98, in SNA, fund Central Committee of Communist Party of Slovakia, box 1 390, fascicle 1 334/19.

28 Summarized concept of environmental care in SSR in relation to the Sixth Five-Year Plan (1976–1980), 11–12, 83–98, in SNA, fund Central Committee of Communist Party of Slovakia, box 1 390, fascicle 1 334/19.
29 Summarized concept of environmental care in SSR in relation to the Sixth Five-Year Plan (1976–1980), 13–16, 99–111, in SNA, fund Central Committee of Communist Party of Slovakia, box 1 390, fascicle 1 334/19.
30 Summarized concept of environmental care in SSR in relation to the Sixth Five-Year Plan (1976–1980), 13–16, 99–111, in SNA, fund Central Committee of Communist Party of Slovakia, box 1 390, fascicle 1 334/19.
31 Summarized concept of environmental care in SSR in relation to the Sixth Five-Year Plan (1976–1980), 16–19, 112–120, in SNA, fund Central Committee of Communist Party of Slovakia, box 1 390, fascicle 1 334/19.
32 Summarized concept of environmental care in SSR in relation to the Sixth Five-Year Plan (1976–1980), 16–19, 112–120, in SNA, fund Central Committee of Communist Party of Slovakia, box 1 390, fascicle 1 334/19.
33 Summarized concept of environmental care in SSR in relation to the Sixth Five-Year Plan (1976–1980), 16–19, 112–120, in SNA, fund Central Committee of Communist Party of Slovakia, box 1 390, fascicle 1 334/19.
34 Summarized concept of environmental care in SSR in relation to the Sixth Five-Year Plan (1976–1980), 121–125, in SNA, fund Central Committee of Communist Party of Slovakia, box 1 390, fascicle 1 334/19.
35 Summarized concept of environmental care in SSR in relation to the Sixth Five-Year Plan (1976–1980), 16–19, 127–135, in SNA, fund Central Committee of Communist Party of Slovakia, box 1 390, fascicle 1 334/19; Mikuláš Huba (ed.), *Slovensko: Slovenské lesy-otázka existenčná (Slovakia: Slovak Forests-Question of Existence)* (Bratislava: Slovenský zväz ochrancov přírody a krajina, 1990), 34–47.
36 Session of Chairmanship of Central Committee of Communist Party of Slovakia, September 9, 1980. Control report about observance of Summarized concept of environmental care in Slovak Socialist Republic and in Bratislava, in SNA, fund Central Committee of Communist Party of Slovakia, box 1 582, fascicle 1 388/13; Ján Budaj (ed.), *Bratislava nahlas (Bratislava Aloud)* (Bratislava, 1987), 1–2; Michal Barnovský (2007), 65–67.

9 Retention of sulfur dioxide emission in the GDR

Between technology, economics, diplomacy, and public opinion

Michel Dupuy (translated from French by Grace Park, University of Chicago)

The environmental history of the GDR from the perspective of air pollution is dominated by problems related to the emission of dust and sulfur dioxide (SO_2) from industrial plants. In 1949, it was technically possible to reduce dust emissions, but as for SO_2, no economically viable solution existed. In 1988, 20,000 km^2 of dust and 48,000 km^2 of SO_2 were released into the East German atmosphere. However, by this time, it was possible to reduce emissions of these two substances, and engineers no longer used the phrase "economically non-viable" when referring to SO_2. Furthermore, the consequences of air pollution for the health of the population and vegetation were well known by the political and scientific authorities. Their costs were estimated at 1.070 million marks, with half of this total from damage done to forest areas.[1] To explain this disastrous situation, several reasons were given: "ton ideology"—in other words, production first rather than quality; the lack of a coherent environmental policy; a lack of finance; an oil crisis; and the aging of the industrial infrastructure. But ultimately, these excuses did not help the East German government to respond effectively, and this failure undermined the regime. By studying the history of air pollution in the GDR at a specific point—the moment when filtration processes against dust and SO_2 emissions were installed—the importance of many actors is demonstrated: individual citizens, engineers, scientists, the political authorities, ecological organizations and international institutions, and shows the limits of effective governance under the GDR government.

Accepting pollution: 1949–1972

Awareness and acceptance of pollution

Pollution in the GDR was related to basic energy resources. In 1945, lignite was the main existing natural energy resource in the GDR, and the government relied evermore on lignite to produce power. In 1946, the GDR extracted 108.4 million tons of brown coal, a number which increased to 137.1 million tons in 1950. In 1953, the 1944 equivalent was reached.[2]

However, lignite has a lower calorific value than other forms of coal, and a higher water content in the range of 50 to 60 percent, and therefore it produces 20 percent less energy when burned in power plants. The sulfur dioxide content varies from area to area. The brown coal from Halle and Leipzig had between 3 and 4.3 percent of SO_2, whereas in Senftenberg (Niederlausitz) that number was between 0.5 and 1.7 percent.[3]

The Forestry Centre of Tharandt (affiliated with the Technical University of Dresden) produced a study of the effects of SO_2 on forest areas in 1953 and re-established a study on air pollution, which had been abandoned in the 1930s. At this time, the concern was more about dust emissions near the power plant than the emission of SO_2, and focused on public hygiene.[4] Thus, in Bitterfeld, in the newspaper *Freiheit* (the organ of the Socialist Unity Party of Germany [SED] in Halle) on October 11, 1953, a letter written by Paul Zänker regretted that "every vehicle, every gust of wind whirls up clouds of flying ash and dust, which, as a result of deficient street cleaning, floats like drifting dunes along the roadsides."[5] Later, on February 2, 1954, in the letters page of *Neues Deutschland* (the official newspaper of the SED), an inhabitant of Berlin named Marianne Schukay complained about the power plant in Klingenberg:

> When we hang out our laundry to dry, it is soon unrecognizable because of the coal dust. Our windows and curtains are always black, as well as our balcony flowers and the fruit in the garden. In the interest of our health, please do something about it.[6]

These complaints were revisited on February 28, 1954, again on the letters page. The answer appeared on March 20, 1954: the power plant would be equipped with filters by 1955.[7] The inhabitants would also need to limit their consumption of electricity during peak periods. Later, on May 26, 1959, *Neue Zeit* announced the construction of a 140-meter high chimney due for completion by 1963.[8] At the same time, in the heart of the chemistry triangle, an East German journalist wrote in July 1963: "man will soon go through Bitterfeld again with open eyes," because the new power plant will be equipped in 1970/1971 with an electro-filter.[9]

With these two examples, the Klingenberg and Bitterfeld power plants, we can find some characteristics that are specific to socialist societies. Most importantly, the population was asked to be patient because the construction of the socialist state required industrialization, a higher priority. In an article from 1965 written by Phil Werr, from the press office of the chemical cooperative of Bitterfeld, the author recalled the importance of agriculture for the cooperative because of the production of fertilizers and pesticide products.[10] Pollution, although acknowledged as a problem, was viewed as a "local norm" (Ortsüblich).[11] Until 1979, in a report on the current situation on the protection of nature in the district of Halle, certain damage was classified as a "local norm." In other words, air pollution was seen as a necessary evil.[12] In such a situation, patience was necessary; in Bitterfeld, citizens would have to wait only eight years, until 1971, as one journalist proudly announced. Other potential solutions included building higher chimneys, or using other energy resources such as gas, oil, or the nuclear option as in other industrialized countries.

Between conscience and inefficiency

As previously mentioned, the forestry center of Tharandt had raised the awareness of the effects of SO_2 on the forest with its studies, reports, articles, exhibitions, and legal actions. The scientists who worked on these topics in Tharandt or in the Institute of Energetics in Leipzig, argued that an "economically viable" process to reduce the emissions of SO_2 did not exist. Thus, a report in 1960 from the Physics/Chemistry Department at the Institute of Energetics claimed that "all processes require substantial investment and operating costs which are not covered by the manufactured product."[13] However, it was only in 1969 that the Institute of Energetics began research on a process for desulfurization. Meanwhile, the only solutions were higher chimneys, different forest species, and compensation payments from companies for damages. Thus in 1965, some foresters from Tharandt evaluated the potential damage of a new power plant on the forest in Niederlausitz. This was estimated at 15 million marks.[14]

Compensation for damages was not common practice for companies at the time. The Institute of Energetics in 1967 remarked that only the Zschornewitz power plant offered damage compensation.[15] Only through legal action could companies be forced to give compensation because companies did not have the right to disturb the production of other companies within the framework of the economic plan. Thus, on July 17, 1967, following a complaint from the forestry district of the Dübener Heide (near Bitterfeld), the Contract Tribunal granted an order against the power plants of Vockerode, Zschornewitz, Muldenstein, the film factory Wolfen and the chemical combine of Bitterfeld.[16] These companies were obliged to provide compensation for losses incurred by the forestry district between 1956 and 1966 on the basis of reports from the Wood and Plant Chemistry Institute of Tharandt.

To avoid these conflicts, the Council of Ministers issued an order on September 8, 1966, relating to the steps to be taken for the protection of air quality. The various ministries involved had to invite companies to modernize their installation facilities. Some standards were issued relating to the emission of dust and SO_2 for longer periods at $0.15\,mg/m^3$ of air. Each company had to comply with these standards and to evaluate dust emission levels. However, they were not equipped with permanent measuring instruments capable of evaluating airborne emissions. These instruments were still under study in the GDR.

In order to measure emissions and to determine the proper compensation, an experiment was conducted in the district of Halle from 1968 to

1970 with the application of a hygiene inspection. Each of 25,000 industrial sources in GDR had to inform an "emission register" sent to the district hygiene inspection, who proposed a compensation scale for the actors who suffered damage as a result of air pollution, including horticulturists, farmers, and foresters. On the basis of this study, under the law relating to nature protection (Landeskulturgesetz) of May 28, 1970, it was decided to entrust these measures to external organizations. The emission from power plants was controlled by the institute of Energetics in Leipzig and the ORGREB-Institute in Vettschau; for the other companies, it was the responsibility of the hygiene inspection in each district.[17] However, the compensation system would only go into effect on January 17, 1973.[18]

The absence of incentives and the lack of pressure ensured that companies and various economic services were not forced to implement measures to reduce pollution, despite the introduction of environmental officers in companies after 1964. The protection of the environment was considered an expenditure. Peter Peklo and Karl Enders (both environmental officers in the Chemiekombinat at Bitterfeld) noticed that their company between 1964 and 1974 devoted 2.21 percent of its expenses to the environment, which they considered to be very little. For them, it was necessary to invest at least an additional 10 percent. However, the situation was critical. At Bitterfeld, 25 percent of the power plant had been built before 1930, and had no filter installed against dust. Twenty-eight percent of the plant had been built between 1931 and 1945 with a *Zyklonentstaubung* (a cyclone-based dust filter) and 47 percent was built after 1945 with electric filtration. However, these were not utilized fully. Thus, in Vockerode, in 1971, only 70.9 percent of the capacity was used, and in Zschornewitz only 20.8 percent.[19]

In 1960, a report written by Wolfgang Kluge at the Institute of Energetics estimated that, in the GDR, power plants "performed their task only in rare cases"; defective installations, misuse, and an absence of sufficient manpower accounted for this situation. In 1967, a new report gave the same results.[20] In cases with more control, emissions nevertheless persisted. Often, companies discharged the dust at night after turning off electric filtration. Moreover, the dust stored was not always removed, and the filtration system could not be activated.[21] As a result, a 1968 report showed that the pollution threshold fixed by the health ministry on August 1, 1966 was exceeded in 50 percent of industrial areas and in 20 percent of areas of population.[22]

The industrial infrastructure was aging, and spare parts were slow to be delivered; the forecasts were not optimistic. In 1967, experts were forecasting an increase in electricity production of 217 percent by 1980, bringing with it a projected increase of SO_2 emissions of 127 percent and dust of 27 percent despite the arrival of oil and the modernization of power plants.[23] In 1976, another forecast written by the Institute of Energetics predicted an increase in air pollution of 8 percent by the year 1990. The author included a section about the process of desulfurization, but he was unable to predict its efficiency because at that time they were still in the prototype stage.[24] In fact, the scientists noted an increase in the emission of pollution. Thus in 1966, power plants in the GDR produced 1.14 million tons of dust (10.5 t km^2) and 1.61 million tons of SO_2 (14.9 t km^2). In 1974, these numbers were 1.4 million tons and 2.12 million tons respectively.[25]

Thus, at all policy levels, from local to national, concerns about the consequences of air pollution were generated by different actors: the general public produced the "Eingabe," or citizen's complaints, the scientists and government officials conducted hygiene inspections, and the Institute of Energetics published their reports. The envisaged solutions were the modernization of the systems of production, the use of alternative energy resources (oil, gas, nuclear) or higher chimneys. It was a slow process, above all, for action against SO_2 emissions.

A new political and diplomatic dimension: 1972–1989

Energy policy reviewed

The beginning of the 1970s was characterized in Eastern Bloc and Western Bloc countries by the growing awareness of environmental questions. In the GDR, it resulted in the creation of a Ministry of Environmental Protection and Water Management on January 1, 1972. The decade ended with an awareness of international issues relating to air pollution with the hole in the ozone layer, which appeared in 1974. These issues appeared in much scientific literature and there was the signing of the "Long-range Transboundary Air Pollution Convention" of 1979.[26] Pollution had entered into the diplomatic context of the Cold War. Moreover, after 1973 the GDR was a member of the United Nations.[27] Thus, it was implicated in international negotiations like these for the writing of the convention of 1979.

The GDR was required to review its energy projections. In 1971, energy trends for the GDR were clear: strongly upward in the use of oil and gas, and above all, nuclear energy would replace brown coal. In fact, the first oil crisis in 1973 served to change that situation. The USSR decided to increase the price of oil, although to a price that was lower than in capitalist countries. Thus, between 1971 and 1978 the price of oil within the COMECON countries increased by 412.1 percent, and within the "non-socialist economic area" by 594.1 percent. After 1974, the GDR reexamined its energy policies. However, Erich Honecker required that supplies of food and consumer goods for the population be fully guaranteed. The GDR always needed more energy because of expected economic growth, but also because households used more and more electrical appliances. The East German leadership did not want to pass on the increased cost of energy to individual families, and it did not want to use inflation. In the end, it was the state budget that bore the cost, leading to an increase in the national debt. In 1981, the USSR delivered less than 3 million tons of coal, and the GDR intensified the exploitation of lignite. In October 1981, the government of the GDR learned that for the year 1982, the USSR decided to reduce the exportation of oil to the GDR from 19 to 16.8 million tons. This was the consequence of the second oil crisis of 1979, but above all because of the poor harvest in the USSR. The state needed to import more cereal from the West and had to pay in dollars. The GDR used the Amsterdam market to procure 1 million tons, and implemented a drastic plan of energy conservation. The conversion of some power plants from oil to brown coal was accelerated. As a result, the domestic consumption of heating oil was reduced from 8.4 million tons in 1979 to 1.9 million tons in 1985, and annual demand was cut by 6.5 million tons. One-third was replaced by solid fuels—mainly raw brown coal—one-quarter by imported natural gas, and another quarter by domestic natural gas and city gas.[28] Some 240 million tons of brown coal were extracted in 1975, 258 in 1980 and 310 in 1988. Nuclear energy could be an alternative, and this was announced in 1957, but by 1985, its contribution was just 4 percent, and 9 percent in 1989.[29]

In this context, SO_2 and dust emissions could not be reduced. In 1987, emissions reached 5.5 million tons of SO_2 and 2.335 of dust.[30] The East German state was unable to ensure a smooth and safe energy transition by using more brown coal and investing in equipment to counter pollution. In 1973, the GDR invested 0.65 percent of its national income in nature protection, but by the end of the 1980s, invested only 0.4 percent.[31]

The limestone additive process: a limited process

In 1969, the GDR began to develop new processes to reduce SO_2 emissions. Among the 200 known processes for the desulfurization of coal, lignite and oil, the Institute of Energetics selected a limestone additive process for wide testing, and requested the department for the protection of the environment and energy to develop the method because the GDR had sufficient underground limestone resources. The process was easy to install, there were no negative consequences for the environment, and no foreign patent was needed.[32] At this time the equation for both the East and the West was the same for a desulfurization process: to develop a low-cost process, cheap to install and maintain, which would require a minimum number of additional operators to be effective.[33]

The first experiment was conducted at the Institute of Energetics in Leipzig, and later at the Dimitrof power plant in Leipzig, for a 64 ton-per-hour steam power application. In addition to the efficiency of this process, experimenters wanted to determine the storage capacities for limestone and the best location to place the plaster that resulted when SO_2 was added to limestone.[34] Added to this were the costs for the company and the running of the electric filter, which worked more extensively. In 1974, the experiment was conducted in a power plant in Lübbenau (220 t/h) always over a short period (between three and six hours).[35] The initial conclusions showed that electricity production was not disturbed and that 40 percent of the SO_2 emissions were retained. In 1981, a pilot application was installed in the power plant of Vockerode (160 t/h).

If this process was developed in the GDR, it was still a prototype in 1984, and it was not suitable for larger power plants. A report from October 1983 noted that for the largest power plant in the GDR at Boxberg, 10,000 tons of limestone would have to be delivered every day, because for three tons of lignite, one ton of limestone was needed.[36] This, clearly, was inconceivable. The proposal encountered other problems. Enders and Peklo confirmed that the addition of limestone created an increase in energy consumption of 10 percent. Above all, a permanent system raised the question of the transportation of the plaster as well as for dust, and the risk was too great to eject the SO_2 into the atmosphere through chimneys.

Installing the limestone process necessitated the building of boiler houses, bringing an extra cost of between 5 and 10 million marks for each power plant. There was more ash, more wear on the electric filter, and there were additional costs for the transportation of limestone and plaster.[37] The heating plant at Schwarzenberg (Karl-Marx Stadt, today Chemnitz) was ultimately equipped with a limestone additive process in 1987. However, the initial budget in 1985 was exceeded by 8.8 million marks because of the production of more ash. Ash evacuation necessitated between fourteen and twenty-five train wagonloads each day. It was possible to reduce this number, but the GDR would have to import from Hungary equipment for the humidification of the ash.[38] The problem with ash concerned the managers at the Max Reimann heating plant in Leipzig, where the limestone additive process was installed. At the moment when workers added the limestone, large quantities of dust were formed, which was then inhaled by the workers. A better filter installation was built on November 30, 1987, but at a cost representing a quarter of the investment of the company.[39] In a report by the Heinrich Hertz Institute (of the Academy of Sciences of the GDR) dated July 2, 1987, the investigator noted that 75 percent of limestone was left unused in the filter because of a failure in the process. Moreover, the ashes were evacuated into landfills, but because of the high degree of acidity, they had to be used again, and the quartz and clay particles in the limestone damaged the electrical filter. In addition, the second time the limestone was used, it retained only 30 percent of the SO_2. Thus, from a technical point of view, the limestone process could not drastically reduce the emissions of SO_2.

An international affair

The convention of 1979 on Trans-Boundary Air Pollution was ratified by the GDR in June 1982 and came into force in 1983. However, when it was ratified, East Germany imposed censorship on all environmental data for all official publications, including press and scientific reviews, because the leadership did not want the data to be used to criticize socialism. Accordingly, this data was classified as a state secret for a period of between five and fifteen years.[40] This coincided with the beginning of the *Waldsterben* (forest decline) affair. Nevertheless, for the West and for its own citizens, the GDR attempted to be a model of pollution control for forest areas. Thus, between 1983 and 1987, several politicians from the West came to visit the GDR: Alfred Dick, in May 1984, Bavarian Minister for development and environmental issues; a Green party delegation in September 1986; and Professor Farthmann (SPD), Minister of Labour, Wealth and Social affairs for the Rhineland Palatinate in November 1987.

These visits were organized according to a supervised schedule. It began with a discussion with Hans Reichelt (Ministry of the Environment). Then there was an excursion to the Dübener Heide (a forest near Bitterfeld), where an afforestation policy was being introduced with resistant tree species. It ended with a visit to the experimental station in Tharandt, where these species were being raised.[41] In 1986, the schedule included the power plant of Vockerode in the North of Bitterfeld, where 40 percent of sulfur emissions were retained by the addition of limestone.[42] The GDR established itself as a model for restoring polluted forest areas and for taking action against SO_2 emissions.

The GDR also wanted to reassure its own citizens. The first article on this process appeared in the national press in 1981.[43] It was in June 1984 that the East German media commented again on this process during the "Conference on the Causes and Prevention of Damage to Forests and Waters by Air Pollution in Europe" in Munich, attended by representatives from the GDR. The Institute of Energetics also received national funding for this process. In articles in the popular press, optimism for the socialist solution was paramount. However, some doubts were expressed in the *Wochenpost* on November 2, 1986, when a journalist concluded that:

> There is not usually limestone as close as it is at Thalheim. And only in a small power plant can we carry out this process often. In a power plant with large steam generators, the continuous transportation of large quantities of limestone must be guaranteed. The same applies for the transportation and application of the plaster.[44]

This doubt was echoed on Radio Rias in West Berlin, which commented several times on this process though its presenters or through such guests as Peter Wensierski who, on June 17, 1985 remarked that if the GDR wanted to reduce SO_2 emissions by 1 million tons, it would be necessary to transport 20 million tons of limestone.[45]

In addition to *Waldsterben*, the GDR was under international pressure because of its adherence to the Long-Range Transboundary Air Pollution Convention ratified in 1982. In June 1984, the GDR aligned with the USSR, the Scandinavian countries, France, and the Federal Republic of Germany

(FRG, or West Germany), a "Concerted Programme for the Reduction of Sulfur Emissions of 30% for the Year 1993" Based on the Year 1980.[46] The East German leadership wanted to prevent the isolation of the GDR from other socialist states and not give support to the notion that the FRG and other capitalist countries were alone in championing nature protection. However, Poland and Czechoslovakia refused to join the treaty because their leaders believed the target to be unattainable. The 1985 Helsinki Protocol on the Reduction of Sulfur Emissions was signed on July 8, 1985, and came into force on September 2, 1987. Rather than reject the treaty, the GDR chose to falsify the data given to the *United Nations Economic Commission for Europe*, the body responsible for monitoring compliance with this convention.

In other respects, the GDR was concerned about the West German Buschhaus power plant, near the border with East Germany. This power plant was planned to run on a high salt content brown coal, and for this brown coal, there was no known desulfurization process. A 300-meter high chimney was built for the dispersion of the pollution. However, this pollution would cross the border and fall on the GDR. The West German Green Party opposed the plant, as did the GDR.[47] Soon in 1983, the East German press published an article about it. As part of the Cooperative Programme for Monitoring and Evaluation of the Long-range Transmission of Air Pollutants in Europe, the GDR installed a third monitoring station in Seehausen in 1983 near Buschhaus in order to measure the flow of SO_2 emissions.[48] The aim was to prove that the GDR imported air pollution. Monitoring at Buschhaus began in 1985, and a filter was expected to be installed in 1987. It was clear that the GDR wanted to be seen publicly as a victim of air pollution from a capitalist country.

Another symbolic location was the Rummelsburg power plant of East Berlin. On February 7, 1985, it was decided to install a Western technology filter to reduce SO_2 emissions. On September 2, 1985, the scientific authorities chose the Wellmann process. On May 29, 1986, a decree from the Council of Ministers confirmed this choice and the British company Davy McKee won the contract, agreeing to reduce SO_2 emissions by 95 percent.[49] The SO_2 was transformed into liquid sulfur (22 kilotons/year) and recycled in the economy, thus avoiding the problems associated with the landfill option.[50] The transportation of the limestone and the liquid sulfur necessitated 2,900 wagons every year and the initial budget was 0.4 million Deutsche marks in 1986. By 1989, it reached 1.3 million Deutsche marks. It was installed on June 25, 1987 but the company encountered some technical problems, especially relating to corrosion. On November 24, 1987, the East German authorities learned via West German TV that the installation of a desulfurization process in Buschhaus did not work, and this was the same company and the same process used at Rummelsburg.[51] In December 1988, the filter was due for operation, but its implementation was postponed to March. Then, the installation again encountered problems with corrosion.

On September 30, 1989, the filter process had to be halted due to a failure in the delivery of ammonia.[52] On September 28, an article in the *Financial Times* about this power plant was translated by the East German ADN Agency, but it was not to be made public and was provided only to government authorities. The article concluded that the late initial entry into operational use was caused by the introduction of new technology from the GDR. The journalist hinted at a struggle for power between the people who wanted to fight against SO_2 emissions and the others. The filter entered operation just before the fall of the Berlin Wall. At the beginning it was designed be a sign of the commitment of the GDR to a policy of reduction of SO_2 emissions, particularly at its borders, but in the end it became a fiasco. It was announced in the East German press and in the West Berlin media, and ultimately demonstrated the weakness of the economy of the GDR.

The use of symbolism was not only the prerogative of the authorities in the GDR. Within the East German ecology movement, place names were used as symbols for pollution. Complaints by inhabitants of the GDR against the effects of emissions were expressed as early as the 1950s from individual citizens. In the 1980s, these complaints still came from individuals but also from people who were more actively engaged in environmental protection.[53] Thus, in October 1983, fifty-four citizens sent a complaint to the Ministry of Environmental Protection and Water Management to protest against the acid rain that was falling on Ores Mountain. The protest was organized by the "Peace and Environment" group in Berlin-Lichtenberg.[54]

After 1978, with the introduction of high school military training, some pacifist groups emerged under the protection of the church. First, they were tolerated by the authorities, because the GDR hoped to show its pacifism on an international scale. In other respects, some ecology groups emerged again under the protection of the church, and in particular with the help of the Church Research Institute in Wittenberg.[55]

Among the most significant actions undertaken in the GDR was the "One Mark for Espenhain" movement. Espenhain was the name of an area with both a power plant and a brickyard, to the south of Leipzig. In July 1987, during a Christian environmental seminar in Mölbis (near Espenhain), organizers invited to participate in the seminar the director of BV Espenhain, the environmental protection officer of the company, and the person responsible for environmental and water management for the district of Leipzig. All these representatives stated that they were not responsible for pollution damage. They did not deny the damage, but they claimed that they were only intermediaries, with the power to make decisions invested at a higher level and there were no hopes for a substantial improvement for the years that followed.[56] During the discussion, the participants learned that the power plant no longer had anti-dust filters. They looked for a solution. In 1988, the Christian environmental group Rötha, with the help of a group from Dresden, initiated the "One Mark for Espenhain" initiative. They knew that it was impossible to buy a dust collector but nevertheless they made a public appeal for donations and launched a petition which 100,000 citizens signed.[57] The "One Mark for Espenhain" initiative was representative. First, it tried to alter public opinion, and second, it promoted the role of the Christian environmental group. They thus created a local space for public debate.

It was difficult by this time to deny the effects of pollution and the requests for patience were no longer tenable. Sometimes in official press releases, reports of non-compliance with the rules about air pollution appeared. In the *Leipziger Volkzeitung* on November 12, 1987, Dieter Halbig, member of the Council for environmental protection and water management, stated that "in the fifty-five cases where specified limits have been exceeded, it was because filter installation was not done in a timely fashion."[58] The companies involved were penalized, but Halbig recognized that there had been some failures.

The implementation of a program: a limited hope

The GDR wanted to show to the international community its sincere intention and readiness to reduce SO_2 emissions by 30 percent. They launched a program in 1985 whereby the GDR finalized its energy transfers. This came just before the signature of the protocol. The "Measures to Recover Resources and to Increase Energy Yield through Flue Gas Desulfurization" was enacted by a decree of the Central Committee of SED on January 15, 1985 and by another decree of the Council of Ministers on February 7, 1985. An inventory of the companies involved in the process installation was prepared. The limestone additive process was planned for three companies in Karl-Marx-Stadt, for the Max Reimann heating plant in Leipzig, and for another plant in Plauen. There was one additional process: the Zittau process (flue gas scrubbing) was proposed for a heating plant to the north of Plauen, but technical problems delayed its installation until June 1988 because of the necessity of importing parts from Western countries. The Chemiekombinat plant at Bitterfeld was obliged to install a wet scrubbing process in 1986, but this was completed only in 1992.[59] On January 10, 1986, the monitoring of these installations was carried out by the "Flue Gas Desulfurization" work group, in collaboration with the Council of Ministers at the beginning of the Ninth Five-Year plan, which was due to stretch from 1986 to 1990.[60]

Simultaneously, following a decree by the Council of Ministers on July 12, 1985, a letter was sent by this work group to all participating parties on July 29, to encourage them to reduce their energy consumption, to find technical solutions, and to encourage recycling. The Ministry for Glass and Ceramic Industries analyzed the answers of the companies under its supervision and replied to the work group on December 18, 1985:

> The available information and materials for flue gas desulfurization is insufficient to enable decision making. In particular for boiler plants with a small capacity (for example from 6.5 tons per hour), economically viable technical solutions are lacking. What was unclear was from which boiler capacity the installation of a flue gas desulfurization made sense [...] There was no qualified consultation from partners in the coal and energy sector.[61]

Despite this political willingness, reality prevailed. The continuing lack of finances was undeniable. Participant parties in the air and refrigeration sectors estimated that East German requirements for the year 1986 roughly amounted to 335 million marks for filter installation, but only 73 million marks were available. In the Ninth Five-Year Plan, only 341 million marks were allotted against a total nationwide demand of 4.273 million marks.[62] For the new desulfurization installations, the GDR needed to produce 200 kilotons of powdered limestone in 1985 and 500 kilotons in 1987. However, the cement production sector simply could not comply with this requirement.

On July 8, 1986, the Ministry of Science and Technology made a critical assessment of the Five Years plan if the GDR hoped to reduce SO_2 emissions by 30 percent as they had agreed. The plan forecast a reduction of 1,518 kilotons in SO_2 emissions. In reality, the reduction would be to the order of 127 kilotons. The lack of investment was fiercely criticized, and companies complained new equipment was needed. The plan therefore provided for forty-five new installations. However, according to the authors of this report, only three were carried out on time, ten were behind schedule, and the last thirty-two were not provided with actual budgetary funds. In order to reduce SO_2 emissions by 127 kilotons, the GDR was obliged to spend 1.5 million Deutsche marks. In order to reduce emissions by 30 percent by 1993, the GDR would have to reduce SO_2 reduction by 2.2 kilotons and this would require the expenditure of between 17 and 20 million marks. In comparison, the FRG reduced its SO_2 emissions by 1.5 million tons at a cost of between 10 and 15 million marks.[63]

The authorities communicated their environmental policies via the media, but on December 9, 1988 a complaint from a private citizen was sent to the primary industry department in the Central Committee concerning the use of the Zittau process in the VEB industrial heat supply plant in Werder/H (District of Potsdam-Land). In the letter, he spoke about the Kirchberger process applied by VEB building management in Zwickau. This process was popularized by the press and on television. One of the reporters explained that the process retained only 50 percent of the SO_2 emissions, and that in addition it released nitrogen oxide, and concluded that "in our estimation, the application of this process is irresponsible. It discredits the policy of environmental protection by the GDR inside and outside its frontiers and these measurement results go against the information provided by the authorities."[64]

The situation regarding air pollution did not improve, but public discontent did not appear to be strong. The lack of public interest was noted in a report to the East German Minister of Environment, Reichelt, in November 1989, concerning the state of the environment in the district of Dresden. The author explained that, despite various commitments undertaken by local authorities, the population was still living in areas suffering from pollution and specified that a political initiative on the part of the local council was particularly difficult and that indeed a great deal had been promised by the Council of Ministers, but in reality nothing had been done.[65]

However, at the end of the 1980s, several new pollution abatement processes emerged: the Dresdner process (developed at the Technical University at Dresden) was applied first in 1985 for 3.5 ton-per-hour steam generators, and the spray absorption process for brown coal power plants was first applied in the Vetschau power plant in 1985. The Dresdner process retained 70 to 80 percent of the SO_2. By means of slaked lime and ashes, the consumption of limestone could be reduced by half, but the process required frequent replacement of the nozzles—and there was always the problem of landfill. This process was scheduled for larger power plants with a capacity of more than 100 tons per hour.[66] The VEB chemical factories in Buna studied the limestone wet scrubbing process for air pollution with the help of three participating organizations: the Institute of Energetics, the Technical University of Merseburg and the VEB dust technology unit, and concluded that the process could be in use by 1994–1995.[67]

Despite the difficulties, political willingness was displayed in some ministries and at the regional level where some agreements had existed between hygiene inspection organizations and the companies involved since the 1960s. However, in May 1989, the district council had to draw up a plan for the reduction of emissions up until 1995. SO_2 emissions decreased a small amount, from 5,500 kilotons in 1987 to 5,208 kilotons in 1988, but they remained above air pollution standards for the GDR.

Internally devised solutions could not reduce SO_2 and dust emissions in the GDR, but on November 13, 1987 during a session of the border commission, the Federal German Republic proposed the installation of a desulfurization process in the power plant because emissions were reaching FRG territory. As of December 1988, the GDR were still refusing this proposition. On July 5, 1989, with the visit of Klaus Töpfer, the federal Minister for the Environment, Nature Protection and Reactor Safety, the leadership of the GDR had a change of heart. On October 23, 1989, Töpfer added Espenhain to his list of projects for the year 1990. He proposed direct financial aid and cheap credit for this environmental project.[68] For the first time, on October 23, 1989, five days after the departure of Honecker, a dispatch from the *Allgemeine Deutsche Nachrichtendienst* referred to emissions in the GDR: 4,990,000 tons of SO_2 and 955,000 tons of nitrogen oxide. This was the first time that this data had appeared, It was the data which was sent to the UNECE for the year 1987, but in fact the real numbers for SO_2 were

5,560,000 tons, and for nitrogen oxide, 820,000.[69] Two East German newspapers reported on it: the *National Zeitung* on October 24 and the *Tribüne* on October 27.[70] The state secret had been revealed and on November 9, 1989, the Berlin Wall fell.

The fall of the Berlin Wall, of course, helped to solve the problem in two ways. First, a reduction of air pollution followed from a decline in industrial production of 66 percent in 1992. Total employment fell from 9.7 to 5.4 million, and the GDP shrunk to half of its pre-1989 level.[71] The effect on energy demand was a sharp fall which continued to 1995. Long-term brown coal production stabilized at around 70 million tons per year. This was much lower than the pre-1989 level of 300 million tons. Second, active power plants were equipped with filters to reduce pollution by SO_2.

Conclusion

The environmental crisis in the GDR and in the other industrial countries in eastern Europe was solved by a brutal transition to a post-Fordist economy and by the modernization of the remaining power plants. Even if the GDR began to implement a policy of reduction for the sulfur emission within various industrial sectors and at the district level, East Germany did not have the capacity to reduce the emissions because of the lack of technology, shortage of economic resources, and an underestimation of the environmental costs. Furthermore, retaining the sulfur dioxide on the ground required a logistics to evacuate it with the dust, but the GDR was unable to achieve it on a large scale.[72] Concerning dust emissions, there were other factors which were more related to the handling of illegal practices (but practices authorized by the company) to release dust into the air by turning off electric filtration processes or by the use of aging or inappropriate equipment.

Pressure from international institutions, from the Federal German Republic, and from ecological movements inside the GDR, coupled with the inability of the GDR authorities to satisfy the demand by its population for consumer goods as in the Federal German Republic, was completed by the fall of the Berlin Wall in 1989. A few years later the environmental crisis was solved but a social crisis emerged.

Notes

1 Arbeitsgruppe für Organisation und Inspektion beim Ministerrat Sekretariat des Ministerrates, Information zur Sicherung des wissenschaftlich-technischen Vorlaufs auf dem Gebiet der Rauchgasentschwefelung, 1986, BArch: DC 20 22680.
2 Falk Küchler, *Die Wirtschaft der DDR. Wirtschaftspolitik und industrielle Rahmenbedingungen 1949 bis 1989*. Berlin: FIDES Verlag, 1997.
3 Wolgang Kluge and E. Wendt, "Staub- und Schwefelgehalt der Abgase von Dampferzeugern mit Staubfeuerungen für Weichbraunkohlen," Institut für Energetik, Hauptabteilung Elektroenergieerzeugung, Bericht Nr.2/1475/66 B, Leipzig, January 25, 1966: 25.

4 Rajanov Svetlozar, *Geschichte der Tharandter Immissionsforschung 1850–2002*. Dresden: Fachrichtung Forstwissenschaften der TU Dresden, 2002; Dupuy, Michel. *Histoire de la pollution atmosphérique en Europe et en RDA*. Paris: L'Harmattan, 2003.
5 Paul Zänker, "Flugasche, eine lästige Begleiterscheinung unserer Industriegebiete," *Freiheit*, October 11, 1953.
6 No name, "Ein Brief aus dem 'Mohrenland,'" *Neues Deutschland*, February 2, 6, 1954.
7 Mo, "Aussprache im Kraftwerk Klingenberg," *Neues Deutschland*, March 20, 1954, 6.
8 Kei, "Neue Schornsteinge für Klingenberg," *Neue Zeit*, May 26, 1959.
9 No name, "Dem Bitterfelder Schmutz den Kampf angesagt," *Freiheit*, July 12, 1963.
10 Phil Werr, "Sorgen mit der Bitterfelder Luft…" *Freiheit*, June 11, 1965.
11 The article 906 of the civil code adopted in 1900 specifies that the pollution in an industrial area is a "local norm" and there were no possible legal proceedings against the factories. Franz-Josef Brueggemeier, *Das unendliche Meer der Lüfte: Luftverschmutzung, Industrialisierung und Risikodebatten im 19. Jahrhundert*. Essen: Klartext, 1996.
12 "Jahresbericht Umweltschutz 1978," February 5, 1979, LA Merseburg: SED-Bezirksleitung Halle, IV D-4 25 050.
13 L. Wolfermann, "Studie zum Problem der Entfernung von Schwefeldioxid aus Rauchgasen, Institut für Energetik," November 1, 1960.
14 Herbert Lux and Günther Flemming, "Über die prognostische Ermittlung der, durch geplante Kraftwerke zu erwartenden Waldbauschäden," Mitteilung der AG Forstl. Rauchschadenforschung Tharandt, 1965, 20.
15 H. Almers, "Konzeption zur Verringerung schädlicher Auswirkungen der Energie- und Kohlebetriebe auf die Umgebung Teilkonzeption: Kraftwerke," October 1967.
16 Michel Dupuy, "Industries, forêts et pollution atmosphérique en RDA 1949–1974," *Histoire, Economie et Société* 21 (2002): 123–134.
17 This law encouraged the firms to respect the threshold of pollution, the political authorities introduced some measures, but at any moment the firms are forced to pay penalties.
18 5. Durchführungs VO zum LKG v. *17.1.1973*, *GBl*. 1 Nr. 18.
19 Karl Enders and Peter Peklo, "Analyse der Verunreinigungen der Luft im Raum Bitterfeld/Wolfen—Istzustand, Auswirkungen, Sanierungsmaßnahmen" (PhD diss., Fakultät für Technische Wissenschaften des Wissenschaftlichen Rates der Bergakademie Freiberg, 1975).
20 Wolfgang Kluge, "Reinhaltung der Luft" Fachtagung vom 21. Bis 23. September 1960 in Wiesbaden, October 29, 1960; Almers, "Konzeption zur Verringerung schädlicher Auswirkungen der Energie- und Kohlebetriebe auf die Umgebung Teilkonzeption: Kraftwerke," 38.
21 Gerhard F. Müller, "Kontrolle der Emission und Einfluss meteorologischer Vorgänge auf die Immission, in Technologie der Abwasserreinigung und Emissionskontrolle der Luft," in Kammer der Technik (ed.), *Technologie der Abwasserreinigung und Emissionskontrolle der Luft*. Leipzig: VEB Deutsches Verlag für Grundstoffindustrie, 1973, 111–130.
22 Dipl.-Ing. Wendt, "Einhaltung von Immissionsgrenzwerten," Institut für Kraftwerke, May 8, 1968.
23 Almers, "Konzeption zur Verringerung schädlicher Auswirkungen der Energie- und Kohlebetriebe auf die Umgebung Teilkonzeption: Kraftwerke."
24 E. Wendt, "Einfluss der langfristigen Energieträgerbilanz auf die Schwefeldioxid-Emission 1976–1990," November 7, 1975.

25 Wolfgang Kluge, "Analyse Umweltschutz," MfKE-Bereich, May 30, 1974.
26 Dupuy, *Histoire de la pollution atmosphérique en Europe et en RDA au XXe siècle*.
27 On September 18, 1973 both countries, the GDR and FRG, joined the UN officially.
28 Felix Christian Matthes, *Stromwirtschaft und deutsche Einheit*. Berlin: F. C. Mathes, 2000.
29 Wolfgang D. Müller, *Geschichte der Kernenergie in der DDR*. Stuttgart: Schäffer-Poeschel, 2001.
30 The emissions of the power plant were about 3.1 million of tons.
31 Jörg Roesler, *Umweltprobleme und Umweltpolitik in der DDR*. Erfurt: LZT, 2006.
32 D. Lippold, "Entwicklungsstand des Kalkstein-Additiv-Verfahren zur Entschwefelung von Rauchgasen braunkohlengefeuerter Dampferzeuger," August 8, 1980.
33 "Berichterstattung der Spezialistendelegation der DDR über die 5. RGW-Arbeitsberatung zum Thema IV.1," May 26, 1982, BArch: DK 5 59.
34 G. Manig, E. Meier, and R. Ungethüm, "Rauchgasentschwefelung," Institut für Energetik, December 12, 1971.
35 Wolfgang Kluge, "Senkung der SO2-Immission in Ballungsgebieten unter Berücksichtigung der Rauchgasentschwefelung durch das Trockenkalksteinverfahren," 1977.
36 Guido Thoms, "Beschluss über die Konzeption für den Einsatz des Kalkstein-Additiv-Verfahrens in der DDR," October 1983, BArch: DK 5 2133.
37 Kuffner, "Information über die Entwicklung der SO2-Emission im Verantwortungsbereich des Ministeriums für Glas- und Keramikindustrie," December 18, 1985, BArch: DG 6 1292.
38 "Arbeitsgruppe für Organisation und Inspektion beim Ministerrat Abteilung Inspektion, Information über den Realisierungsstand der Einsatz des Kalkstein-Additiv-Verfahrens in der DDR zur Minderung der Schwefeldioxidemission," Mai 1985, BArch: DC 20 11182.
39 Ribbecke, "Information über Probleme auf dem Gebiet der Rauchgasentschwefelung," October 26, 1987, BArch: DH 1 36036.
40 Michel Dupuy, "Science, pouvoir et pluies acides en RDA," in L. Coumel, R. Morera, and A. Vrignon (eds.), *Pouvoirs et Environnement. Entre confiance et défiance, XVe–XXIe siècle*. Rennes: Presses Universitaires de Rennes, 2017, 159–173.
41 "Weiterführung der informativen Gespräche zwischen Experten der DDR und der BRD zu Fragen des Umweltschutzes, insbesondere zur Rauchgasentschwefelung von Kohlekraftwerken," August 11, 1983, SAPMO-BArch: DY 30 2586.
42 This process was piloted by the Institute of Energetic in Leipzig in 1985. It was the result of a program launched in 1966. Wolgang Kluge, "Konzeption zur Verringerung schädlicher Auswirkungen der Energie- und Kohlebetriebe auf die Umgebung Teilkonzeption: Kraftwerke," IFE-Bericht Nr. 2/1763/67 F, 1967.
43 Jürgen Nowak, "Nun weniger Schwefel durch den Schornstein," *Neues Deutschland*, August 15, 1981.
44 Ina Pachmann, "Ein Prototyp überraschte angenehm," *Wochenpost*, November 2, 1986.
45 Peter Wensierski was a West German journalist, who work on the GDR and in particularly on the environmental movement. "Der Wald stirbt nicht—Er verändert sich"—Umweltschutz in der DDR. Im Landfunk hören Sie heute einen Beitrag von Peter Wensierksi, Rias, June 17, 1985, Rias Archiv: 9588.
46 Reduction of 30 percent of the flux did not imply a reduction in emissions to the same extent. This principle had already been defended by USSR before the signing of the 1979 convention. Evgeny Chossudovsky, *"East-West" Diplomacy for Environment in the United Nations: The High-Level Meeting within the Framework of the ECE on the Protection of the Environment*. New York: UNITAR, 1990.

47 No name, "Buschhaus: Da heften sich menschliche Gefühle an einen Popanz," *Die Zeit*, August 3, 1984.
48 No name, "Unvertretbare Umweltbelastung durch Kraftwerk in der BRD," *Neues Deutschland*, June 25, 1984.
49 It was the same firm, which equipped the filter installation in the power plant of Buschhaus in FGR.
50 "Beschluss des Politbüros des ZK der SED vom 27. Mai 1986," BArch: DC 20 I 4 5807.
51 Dr. Kratzke, "Information zu Meldungen des BRD-Fernsehens zum Kraftwerk Buschhaus am 24 November 1987," BArch: DG 12 1519.
52 Sedler, "Volkseigener Aussenhandelsbetrieb Industrieanalage Import, Brief to Mr Tolimson, Davy McKee," October 1, 1989, BArch: DL 2 16673.
53 Felix Mühlberg, "Informelle Konfliktbewältigung. Zur Geschichte der Eingabe in der DDR," PhD diss., Technischen Universität Chemnitz, 1999.
54 Tobias Huff, *Natur und Industrie im Sozialismus. Eine Umweltgeschichte der DDR*. Göttingen: Vandenhoeck & Ruprecht, 2015.
55 Hans-Joachim Veen, *Lexikon Opposition und Widerstand in der SED-Diktatur*. Berlin-München: Propyläen, 2000.
56 Roland Quester, "5. Umweltgottesdienst," *Streiflichter* July, 1987.
57 Karl Weise, interview by Michel Dupuy, *Pegau*, July 24, 2000.
58 Dieter Halbig, "Weitere Maβnahmen für bessere Leipziger Luft?," *Leipziger Volkszeitung*, November 12, 1987.
59 "Information zur Sicherung des wissenschaftlich-technischen Vorlaufs auf dem Gebiet der Rauchgasentschwefelung," 1986, BArch: DC 20 22680.
60 Willi Stoph, "Beschluss über die Bildung der Arbeitsgruppe des Ministerrates zur Rauchgasentschwefelung," January 10, 1986, BArch: DC 20-I-3 2270.
61 Kuffner, "Information über die Entwicklung der SO2-Emission im Verantwortungsbereich des Ministeriums für Glas- und Keramikindustrie," December 18, 1985, BArch: DG 6 1292.
62 BArch: DC 20 22680, op. cit.
63 Dr. F. Krause, "Stellungnahme zur Vorlage von Gen. Dr. Götz Einordnung der Vorhaben der Rauchgasentschwefelung im Fünfjahrplan 1986–1990," July 8, 1986, BArch: DF 4 22553.
64 Wolgang Kluge Seibt, "Bericht über die Untersuchung zum Sachverhalt der Eingabe des Bürgers Stefan Barth vom 9.12.1988 and die Abteilung Grundstoffindustrie beim ZK der SED zur Anwendung des Zittauer Verfahrens zur Abwärmenutzung und Rauchgasentschwefelung im VEB Industrielle Wärmeversorgung Werder/H/," February 10, 1989, BArch: DC 14 2512.
65 Witteck, "Konzeption zur schnellen und spürbaren Verbesserung der Umweltbedingungen in den Ballungsgebieten des Bezirkes Dresden," November 1989, BArch: DK 5 3386.
66 Reise, "Protokoll über die 6. Beratung der Arbeitsgruppe des Ministerrates zur Rauchgasentschwefelung am 18 November 1987," BArch: DH 1 30036.
67 Kommission Reinhaltung der Luft, "Einschätzung des aktuellen Standes der Forschungsarbeiten zur Rauchgasentschwefelung," February 10, 1985, BArch: DF 4 22553.
68 SAPMO-BArch: DY 30 3134.
69 "Übergabe von Daten und Informationen über Luftverunreinigungen der DDR an die Wirtschaftskommission der Vereinten Nationen für Europa (ECE) und den Rat für Gegenseitige Wirtschaftshilfe," June 28, 1988, BArch: DK 5 2115.
70 Rolf Bartomek, "Schaden unsere Preise der Umwelt?" *Nationalzeitung*, October 24, 1989; Rolf Bartonek, "Exakte Daten aus dem chemischen Hexenkessel," *Tribüne*, October 17, 1989.
71 Matthes, *Stromwirtschaft und deutsche Einheit*.

72 In 1988, the debt of the GDR raised 123 billions of marks, more than half of its budget for 1988. Steiner André, "From Soviet Occupation Zone to 'New Eastern States.' A Survey," in H. Berghoff and U. A. Balbier (eds.), *The East German Economy, 1945–2010, Falling Behind or Catching Up?* Cambridge: Cambridge University Press, 2013, 17–51.

10 Sino-silviculture

State-sponsored green forestry initiatives in Mao's China

Christopher Reed

"The trees are like my children." That was the answer I got, over tea, from the sagacious gentleman sitting across from me. The question? I had asked him how he viewed China's forestry initiatives in the late 1950s. My partner for tea that July afternoon was a government worker from Chengdu who held a low-level party position with the local forestry division. Our interview was scheduled for one o'clock in the afternoon on that hazy, sun-stifled day in Chengdu's industrial region. On two different occasions that morning he had telephoned to cancel the meeting. He was nervous of reprisals by his superiors should they find out he was talking to a foreigner about government business. After several assurances that I was a scholar of history and not a journalist, along with my vow to protect his anonymity, he finally agreed to meet.[1]

"Lao Lin" (or "Old Trees"), as I'll call him, sat down for our discussion while my guide poured tea from the electric kettle into tiny ceramic cups. Lao Lin was no different from many of his fellow septuagenarians—short and frail—the stature typical of Chinese from that era, evidence of malnutrition via a hungry upbringing during the late civil war era. He was still visibly nervous as he sunk into the Western-style sofa across from me. He was dressed in shorts, a t-shirt, and sandals, and his knee never quit bouncing the whole time we talked. He also fiddled continuously with the large red-beaded good-luck bracelet he wore on his wrist. Nerves notwithstanding, he was always smiling and his eyes lit up when we talked about the trees—or "his trees" as he liked to call them. At one point, he confessed that "we didn't know any better" when we were talking about the very early years of the People's Republic of China (PRC) and their nascent environmental policies.

Lao Lin grew up in Luding (Canzi prefecture) with his parents and five brothers and sisters. He was the first in his family to attend high school. In school, he was fascinated by forestry and his first job out of high school was planting and transplanting trees for the government in the mountainous regions. He grew up poor and could not recall anyone in his village who was not in the same predicament. "People just needed to survive," he said. In explaining the planned economy the way he witnessed it, he simply said that "people had to work more to score more work points to get more food." As he sat across from me, I could sense his rumination when reflecting on his youth when he said, "it's hard for people to understand 'endangering the environment' versus eating." When our interview ended, on his way out the door, he turned to me and left me with some parting insight: "having a family to feed does not afford one the luxury of a consciousness toward the environment"—poignant words from a man proud of his years of service in Chinese silviculture. It gave me pause to stop and consider the weight of his septuagenarian wisdom. This concept of not having a "luxury of consciousness," or not understanding environmental damage versus eating could just as well be applied to a patriarchal Mao Zedong, who himself had not just one family but an entire nation to feed and grow. Did this lack of luxury and understanding drive China to the careless ends-justify-means measures against the environment that the current historiography suggests?

The major goal of this chapter is to reconsider the prevailing scientific consensus on the relationship between the natural environment and the Chinese Communist Party under Mao Zedong. That is, Mao's rapacious pursuit of agricultural and industrial modernization with the complete and deliberate disregard for nature conservation. Given China's complex and uncharted environmental history, the complex assessment of the most populated country in the world cannot be the goal of this humble essay. What this chapter may examine, however, is the desire to protect the environment from needless destruction in communist China, even in the period of strictest central party control. This chapter is not a hagiography of Mao Zedong, but neither is it a castigation of a corrupt and morally bankrupt regime as the current historiography suggests. Rather, this modest chapter aims to serve as a surveillance on the existing literature on modern Chinese environmental history with the quest to seek sparks of conservation and seed ideas of environmentalism under Mao Zedong. To complement the literature survey I will use newspaper articles from 1957 and 1958, at the beginning of China's Second Five-Year Plan, and the time when the Party launched the Great Leap Forward (GLF).[2] As Frank Dikotter writes in *Mao's Great Famine*, any attempt to understand what happened in communist China must start by searching "squarely at the center" of the Maoist period. He argues that the "pivotal event" in the history of the PRC was the Great Leap Forward. Dikotter also adds that Mao was the "key architect" of the GLF, and thus "bears the main responsibility."[3] Much of the current historiography uses the beginning of the Great Leap Forward as the onset of "coercive state behavior [that] ... contributed directly to a wide range of environmental problems."[4]

Some historians suggest that "as the modern world struggles to find a balance between freedom and regulation, the catastrophe unleashed [under Mao] stands as a reminder of how misplaced is the idea of state planning as an anecdote to chaos," but this chapter will attempt to answer a new set of questions using the primary sources.[5] This chapter will demonstrate that Mao Zedong and the CCP did try to protect the environment, and even enhance it, during the Great Leap Forward. This chapter will further argue that when compared with economic development in non-authoritarian, Western nations, the only marked difference in China's record is one of space and time. In other words, the era in which China's industrialization occurred and the lens through which it was observed produces the perception that China's environmental record was worse than that of the West before it.

The current historiographical paradigm

There is a historiographical paradigm that tends, in the words of Peter Ho, "to sweep the mass campaigns into one heap," and lay blame for environmental problems with Mao Zedong and the CCP.[6] For instance, in his work about environmental sustainability in rural China, Bryan Tilt is predominantly concerned with current-day China's environmental concerns.[7] However, to get to the root of contemporary China's problems, he also discusses the Maoist era, and he does not hold back in his castigation of Mao Zedong's tactics. Among several of Tilt's claims is that China's period of high socialism was marked by severe environmental degradation, including widespread deforestation. He also contends that China's rich philosophical tradition of "harmony between people and nature" was effectively displaced from the collective consciousness by Mao's "spasmodic campaigns." Of course, Tilt is not the only scholar working within the current paradigm; in his own words, Tilt says, "the most accomplished scholar of the Chinese environment" is Vaclav Smil.[8] Vaclav Smil has written extensively about the environmental conditions in China. In *The Bad Earth*, his aim is to focus on many degradative trends to which the Chinese environment has been subjected."[9] Unsurprisingly, Smil and Tilt make many of the same claims about the Maoist era. Where Tilt finds "spasmodic" episodes, Smil argues that "a greater ecological lunacy is hard to imagine."[10] Another scholar of Chinese environmental history, Robert Marks, focuses his attention on the Maoist era and the CCP's attitude toward forests. Marks claims that the state announced that forests were "an inefficient land use," and that trees were either to be "exploited or moved out of the way" for the production of grain.[11] Of all the Chinese environmental historians, however, Judith Shapiro makes perhaps the strongest argument against Mao and his regime. Shapiro states Mao created an environment which imposed upon the people a "lack of freedom to behave responsibly toward the natural world."[12] Taken together, Tilt, Smil, Marks, and Shapiro lead readers to believe that Mao Zedong was a puppetmaster in an anti-environmental revolution and 700 million minions willingly did his bidding to dam the streams, turn swamps into cropland, and cut down every tree with reckless abandon, in exchange for an agricultural and industrial "leap" that allowed them to make steel in backyard forges that ultimately proved useless.[13]

Evaluating the paradigm

There is ample evidence from primary documents undermining the claims cited above. Tilt's claims about environmental degradation and state indifference, for instance, are not corroborated by sources from the period. In refutation of the claim of widespread deforestation, for instance, the New China News Agency reported on January 3, 1958, that 10,000,000 hectares of land was afforested under China's First Five-Year Plan between 1953 and 1957. The article noted that this was "more than the total afforestation areas of Britain, Belgium, the Netherlands, Greece and Italy."[14] Tilt's claim that China's move away from a harmonic relationship with nature was due to Mao's "spasmodic" campaigns—that is, irrational and illogical decisions—is contradicted by sources suggesting a calm, rational approach and planning to the environment. To illustrate planning, the sources announced that "in a period of twenty years, we shall basically complete the afforestation of all our mountainous regions." To the claim of "spasmodism," the sources show a government looking to the future with careful planning for their forests: "Avoid logging without planting new trees" was the order of the day. An irrational regime would never admit "[t]here are at present serious cases of timber waste," or vow to "make rational use of forests."[15]

To Smil's charge of lunacy, an article from December 8, 1957 made several proclamations that do not sound like those on the lunatic fringe. The Party reported that their mission was "to take good care of the available forest and make rational use of them side by side with an effort to build and nurse more." They continued by stating that forestry should be planned and carried out in accordance with the principle of "overall planning, comprehensive management, and centralized control."[16] Although Smil claims that in deforestation, soil erosion, and losses of arable land, the country's poor record rivals that of the worst afflicted nations in the world," the sources show careful consideration about forest protection and planting.[17] An article from November 28, 1957 reports that a national forestry plan was implemented to "plant as many trees" as possible, because only with an adequate number of forests was it possible to "conserve soil and cultivate local products."[18] Further, the article proclaimed that "we must not take light of forest protection" and that "afforestation is a long-term project," while asking people to "avoid logging without planting new trees."[19] Smil further claims that, in addition to "resource exploitation by the state," there was rampant "desperate actions of poor peasants."[20] On January 2, 1958, the state reported that a large number of trees had been planted in the form of shelterbelts in the fight against erosion.[21] Personal testimony also helps understand the environmental attitudes of the Great Leap Forward era. Lao Lin's recollections refute some of the claims about desperate, poor peasants. He remembered that everyone in his village was in the same predicament, yet he was unable to recall anyone who partook in illegal cutting or pirating the local forests.[22]

The claims made by Marks about prevailing state attitudes about forested land are likewise challenged by the historical record. An article from the New China News Agency from November 28, 1957, for example, conveyed the Party's directions to the people, starting out by admitting that "in recent years, we have felled quite a number of trees without building enough to replace them," and then imploring the masses "to develop forestry by building, nursing, and protecting trees." The state also said that "we should build a forest land of 800,000,000 mou within ten years" and continued on to say that this plan was "very good and should be carried out." The Party made a strong case against clear-cutting areas by stating that "when we log a large state-owned forest, let us avoid making a 'clean-shave' of it ... and leave the young to grow."[23] Articles published the next month show even more Party concern for forests. It was reported that the "protection of forests ... will have an important bearing on production," and that "we should see to it that state farms also undertake a proper proportion of afforestation."[24] Then, on December 28, the state announced that Kiangsu Province had planted 40 million trees and afforested 12,000 hectares.[25] In direct refute of the claim that the Party thought trees were an "inefficient land use," the Party stated that cooperatives should "plant timber, special industrial and firewood trees," and see to it that long-term afforestation be coordinated with short-term crop production "so that this may serve that."[26] Clearly, the state wanted to coordinate long-term afforestation with crop production so as to ensure a symbiotic silviculture and agriculture relationship, and did not view trees as "inefficient" or antagonistic to crop production.

As for Shapiro's claim that Mao eliminated the freedom to interact responsibly with the natural world, it should be remembered that average people like Lao Lin were out there—everyday people who loved their jobs in the forest and came to feel that the "trees were like my children."[27] Lao Lin remembered that wood was the main economic material in China in the 1950s, and that China depended on wood to survive. However, he reminded me, after 1958, new forestry initiatives began and today "we have trees this big"—holding his hands about two feet apart like a fisherman bragging about his catch.[28] Obviously, there was not a complete lack of freedom when it came to the average person interacting with their environment.

Primary sources also illustrate the degree to which CCP's environmentally related policies represented improvements over its imperial predecessor. One article stated that the total area planted with trees was "thirty-three times" the total planted under the "entire Guomindang regime in the thirty-five years before Chinese liberation."[29] Another article, related to the education of forestry experts, contradicts the often-encountered assertion about Mao's rejection of the use of experts and the CCP's nonchalant approach to environmental concerns and forest management. This article, entitled "Education on Forestry Develops Fast in Our Country," began by stating that forestry work in China "has had such an unprecedented development that it completely frustrates the slanderous charge that this work has been a mess." The article went on to say:

> In the sixteen-year period [leading up] to 1946 of old China, schools of higher learning graduated on average fifty-one students of forestry per year, more than two-thirds of whom either became unemployed or worked on some other posts. The forestry education has had a fast development since liberation. In 1952, a forestry college was established in Peking, Nanking, and Harbin, and forestry departments were set up in thirteen of the schools of higher learning. Students in forestry institutes reached five to six thousand ... representing seventeen times the average number of graduates prior to liberation.

In addition to forestry graduates, the article also boasted that, in the seven years to 1956, the nation established forests representing forty times the total area planted in the thirty-five years prior to liberation.[30]

Challenging the paradigm

Given the examples and evidence thus far regarding Mao's regime and its predecessors, one is led to ask: why do Chinese environmental historians usually castigate Mao and accuse him of waging a war against nature? One scholar who suggests an answer to this question is Peter Ho.[31] While Ho's work centers not on environmental issues per se, but rather on the Grain-First campaign in the period of collectivism from 1956 to 1978, his work argues for a new look at sources, because, as Ho claims, "Chinese and western authors have recorded the period of collectivization *solely* in negative terms."[32] Ho initially set out to research increased desertification resulting from the baleful decisions during the GLF and the Grain-first campaign. However, as Ho explains, while doing his archival research he "stumbled upon a number of inconsistencies that cannot be explained from the [current] view of the period."[33] Ho investigated the key slogan of the dominant discourse, "Take Grain as the Key Link" (yi liang wei gang). While the predominant interpretation tells us this slogan was the basis for "indiscriminate, large-scale agricultural reclamation that took a heavy toll on the environment," Ho's interpretation complains that "a powerful discourse surrounds the topic that filters out contradiction of dissenting voices."[34] Ho, in his own words, aspires to demonstrate that the difference between the primary and secondary sources is the result of recent source selection to "construct a disparaging retrospective discourse."[35] Ho's research further explains that the above quote often repeated by historians is only one portion of the actual full quote used by the Chinese government. The actual quote states: "Take Grain as the Key Link for Overall Development and Diversification" (*yi liang wei gang*, quanmian fazhan, *duoxhong jingying*). Ho also demonstrates an important facet that is altogether missed or neglected by most historians, another slogan that Ho insists accompanied the one mentioned above: "Agriculture, forestry, and animal husbandry are interdependent; not one can be omitted."[36] Ho's conclusion is that there has been an "exclusion, or juggling" of statistical data so as to "support a[n] inaccurate reading of the Maoist era." Since the 1970s, he adds, "this discourse gained ascendancy and evolved into the dominant narrative that exists today." He further contends that reinvestigating the sources, or a renewed investigation of the textual sources written during the time, should "deter us from uncritically accepting the narrative" of the current historiography.[37]

Other documents lend credence to Ho's argument that primary sources have been interpreted in tendentious ways. In an article published on December 8, 1957 in Beijing, titled "Bright Future for Development of Mountain Economy," the author repeats no fewer than three times the need to "lay stress on a variegated management that brings together agriculture, forestry, animal husbandry … and make them administer to each other's needs to push each other forward."[38] This article echoes another from three weeks earlier, which asserted that "mountainous production is multifarious, embracing forestry, [and] animal husbandry."[39]

The paradigm in context

Prior to the October 1949 victory of the Chinese Communist Party (CCP) over their rivals, the Communists promised the people a path to modernity and prosperity in exchange for their loyalty and adherence to initiatives. Because of poverty, technological backwardness, and international isolation, the Chinese had few options but to focus on developing their economy as rapidly as possible. Of course, that meant the natural environment would be required to play a part in that development. Despite this, China's industrialization was no worse for the environment than other nations during their own time of industrialization.[40] Direct comparisons provide important context.

Europeans developed modern silvicultural practices. It is important, as a means of comparison, to explore the European silvicultural record. A popular textbook on Western Civilization gives an account of European deforestation and agricultural practices that will sound eerily familiar to anyone who has read Shapiro, Marks, or Smil:

> vast tracks of virgin forests were cleared for farming, especially in eastern Europe. After all the good bottomland was cleared, farmers moved to clear the more marginal land on hills and mountainsides. These clearings created soil erosion that contributed to devastating floods. Thus, human actions facilitated the ecological catastrophes. No more virgin land was available for clearing, which meant that a

> still-growing population tried to survive on a fixed amount of farming land. The ability of farmers to produce food could not keep up with unchecked population growth. The propensity for famine was especially acute in heavily populated western Europe.[41]

The above reads like an excerpt straight from the pages of *Mao's War Against Nature*, but it is actually about a Western society not under the repressive thumb of a dictatorial regime.

Scholars of Chinese environmental history often provide statistical data to support their claims, but usually omit comparisons that provide important context. For example, Vaclav Smil provides specific figures about the environmental toll in communist China; he states that the Chinese clear-cut 3.29 million acres of forest during Mao's first ten years.[42] Meanwhile, in the United States, the height of the campaign to rampantly industrialize began in earnest in the mid-to-late nineteenth century. Using the dates of 1870–1880 as the beginning of a "new era marked by a determined state effort to industrialize," we find a ten-year period where the United States also deforested in earnest. During the exact same amount of time which Smil lays out China's clear-cutting zeal, the United States denuded over 30 million acres of forest: nearly ten times as much as did the Chinese during Mao's first decade. Robert Marks claims that Chinese deforestation not only degraded the environment but also brought on shortages of timber and wood, critical issues in an agrarian economy still largely fueled by tapping energy stored in trees.[43] Here, Marks may have been discussing China, but could have been writing about the United States. Even as late as 1909, the estimated "growth and drain" ratio in the United States forestland was one to three-and-a-half. In other words, even in the twentieth century the United States was deforesting three-and-a-half acres of timber for every acre it planted. In 1933 alone, the United States planted 8.9 billion cubic feet of forest volume while deforesting over fourteen-and-one-half billion cubic feet. In fact, it was not until as late as 1962 that the United States broke equilibrium on the growth and drain ratio and began consuming .97 acres for every one that was planted.[44] In 1952 and again in 1956 the United States far outpaced the Chinese in deforestation (see Table 10.1 below), yet this data is wholly neglected or, in Ho's words, there is an "exclusion or juggling" of the data related to Mao's China. The argument for this phenomenon could be one of space and time. On one hand—space—you have the propensity of "othering." Historians can be just as guilty as anyone

Table 10.1 Timber area deforested in China and the USA[45]

Year	*China (vol. in cubic feet)*	*USA (vol. in cubic feet)*
1952	229,554,334	10,780,000,000
1956	4,979,368,008	11,280,000,000

in having a slightly easier time finding faults in the way other cultures achieve their ends. The second—the time aspect—is the more plausible reason. While the United States cut down nearly ten times the timberland over one decade during the process of industrialization, it occurred in a time of relative media absence. In the late nineteenth century, significantly less people—outside of the Bureau of Census Land Clearing department in Washington, DC—were concerned, or were even notified about deforestation. However, China during the 1950s was under the world's microscope. Literally, daily reports were circulated globally about every move Mao and the CCP made. The whole world was watching China's development in the 1950s, almost no one was scrutinizing the U.S. development during its industrialization infancy. (Even fewer were watching Europe during theirs.)

Conclusion

Throughout history, every modern nation-state has carried out some form of deforestation in order to develop its economy. Authoritarian regimes such as Mao's China, however, are often criticized for especially predatory forestry practices, so destructive as to leave their nations nearly depleted of vegetation. A closer look at Chinese practices shows this perception to be exaggerated if not mistaken. "The trees are like my children" is a sentimental rumination one would not expect to hear from one of Mao's minions working in the forestry industry during the GLF era, and yet this is how he remembers his work. Moreover, Lao Lin's affection for his trees reached up to the top of the political structure. On December 8, 1957, an article appeared declaring that "the current mission is to take good care of the available forests and make rational use of them side by side with an effort to build and nurse more."[46] China was under the world's microscope during the 1950s, with many secretly (and some not so secretly) hoping to see them stumble under the new leadership. Addressing the naysayers of the day, and perhaps anticipating Shapiro's claim that "only after Mao's death were [forest workers] permitted to reforest the mountains,"[47] a 1957 report stated that "forestry work in our country has had such an unprecedented development that it completely frustrates the slanderous charge that this work has been in a mess."[48] This evidence suggests that, current historiography notwithstanding, Mao's environmental decisions did not oppose care for the natural world, and they included important actions that attempted to foster environmental care and appeared to create a "green dictatorship" in China.

Notes

1 "Lao Lin," 2016. Interview by author (with assistance from interpreter for Sichuan dialect. Chengdu, China, July 9. Lao Lin is a pseudonym chosen to keep the promise to protect the anonymity of my source.

2 Newspaper articles are derived from *Survey of China Mainland Press* from the American Consulate General in Hong Kong in 1958. American Consulate

workers obtained and translated Chinese newspapers from throughout mainland China and sent the translated manuscripts back to Washington, DC as a means of showing the State Department what was going on in mainland China. The following disclaimer is typewritten at the bottom of each table of contents: "Unless otherwise indicated, these materials are full translations of articles appearing in Chinese Communist newspapers or releases emanating from Chinese Communist news agencies. They should not be construed as representing views of the United States Government."

3 Frank Dikotter, *Mao's Great Famine: History of China's Most Devastating Catastrophe, 1958–1962* (London: Bloomsbury Publishing, 2010), xii, xiii.
4 Judith Shapiro, *Mao's War Against Nature: Politics and the Environment in Revolutionary China* (New York: Cambridge University Press, 2001), xii.
5 Dikotter, *Mao's Great Famine*, xii.
6 Peter Ho, "Mao's War Against Nature? The Environmental Impact of the Grain-First Campaign in China," *The China Journal* 50 (July 2003): 40.
7 Bryan Tilt, *The Struggle for Sustainability in Rural China: Environmental Values and Civil Society* (New York: Columbia University Press, 2010).
8 Tilt, *Struggle for Sustainability*, 84.
9 Vaclav Smil, *The Bad Earth: Environmental Degradation in China* (Armonk: M. E. Sharpe, 1984), 4.
10 Ibid., 21.
11 Robert Marks, *China: Its Environment and History* (Lanham, MD: Rowman and Littlefield Publishers, Inc., 2012), 286.
12 Ibid., 195.
13 Shapiro, *Mao's War Against Nature*, 75.
14 *Survey Chinese Mainland Press*, American Consulate, Hong Kong. January 3, 1958. No. 1686, 28.
15 Ibid., November 28, 1957. No. 1670, 32.
16 *South China Morning Post* (SCMP), December 8, 1957. No. 1683, 8.
17 Smil, *The Bad Earth*, xii.
18 SCMP, November 28, 1957. No. 1670, 31–32.
19 Ibid., 31–32.
20 Smil, *The Bad Earth*, 9.
21 SCMP, January 2, 1958. No. 1686, 2.
22 Interview, *Lao Lin*. It should be noted that the question about pirating of local trees was directly asked during the interview due to my reading of the current historiography. I was under the impression this was normal and I wanted his opinion and insight of those events.
23 SCMP, November 28, 1957. No. 1670, 31–32.
24 Ibid., December 8, 1957. No. 1683, 8–9.
25 Ibid., December 28, 1957. No. 1682, 18.
26 Ibid., December 8, 1957. No. 1683, 8.
27 Interview, *Lao Lin*.
28 Ibid.
29 SCMP, January 3, 1958. No. 1686, 28.
30 Ibid., October 13, 1957. No. 1642, 20.
31 Ho, *Mao's War Against Nature?* 37–59.
32 Ibid., 38 (emphasis is author's not mine).
33 Ibid., 40.
34 Ibid., 39.
35 Ibid., 41.
36 Ibid., 50.
37 Ho, *Mao's War Against Nature?* 58, 40.
38 SCMP, No. 1683, December 8, 1957, 8.

39 SCMP, No. 1683, December 8, 1957, 8. No. 1670, November 28, 1957, 30.
40 Marks, *China*, 270.
41 Brian Levack, Edward Muir, and Meredith Veldman, *The West: Encounters and Transformations*, 5th edition (Boston, MA: Pearson Education Inc., 2017), 284.
42 Smil, *The Bad Earth*, 15.
43 Ibid., 251, 271.
44 National Report on Forest Resources: www.fia.fs.fed.us/slides/major-trends.pdf. Accessed: November 22, 2017. Also, US Forest Cover: https://ourworldindata.org/forest-cover/#annual-net-growth-and-use-of-timber-in-the-united-states-1800-2040-williams-2006ref. Accessed: November 15, 2016.
45 For United States: U.S. Timber Production, Trade, Consumption, and Price Statistics 1950–1980. http://archive.org/stream/ustimberproducti1408ulri#page/6/mode/2up. Accessed: November 22, 2017. For China: SCMP, September 25, 1957. No. 1624, 19.
46 SCMP, December 8, 1957. No. 1683, 8.
47 Shapiro, *Mao's War Against Nature*, 84.
48 Ibid., October 13, 1957. No. 1642, 20.

11 Zimbabwe's drift toward authoritarianism and its environmental consequences, 2000–2017[1]

Muchaparara Musemwa

Introduction

On September 16, 1980, barely six months after Zimbabwe's attainment of independence from the Rhodesian white minority rule of Ian Smith, the new Prime Minister, Robert Gabriel Mugabe, was invited to give the keynote speech at the *First National Conservation Congress of the Natural Resources Board (NRB) of Zimbabwe*. In his address, Mugabe dispelled the heightened anxieties within the white community, which had presumed what his Marxist–Leninist government could possibly do to the white populace let alone to a colonial institution such as the NRB, which had championed the conservation of environmental resources since the 1940s. Mugabe surprised his predominantly white audience with his reconciliatory tone when he remarked,

> There is a great deal of anxiety within the Conservation Movement … I am not sure why this should be because they would surely be a stupid people today who did not obey the laws of nature and conservation, particularly what has happened in the history of man elsewhere in the world.[2]

Mugabe's message served two purposes. The first was that he sought to assuage the fears of a restless white minority which had rightly or wrongly been made to believe, through Rhodesian propaganda, that the professed Marxist–Leninist-aligned leader and his government would annihilate or expel them from the new Zimbabwe upon his assumption of power.[3] The second intention was to placate the fears of white environmentalists who had, by the time of independence, built a formidable environmental conservation governance institution—the Natural Resources Board.[4] This body had succeeded in demonstrating that the colony's resource-dependent economic development strategy was unsustainable.

Mugabe had earlier made these harmonious overtures on March 4, 1980 in his celebrated national reconciliation speech marking the transfer of power from Britain to the new Zimbabwe government in order to dispel white fears about the possibility of revenge from the Black government. Yet, as some scholars have argued, Mugabe's conciliatory speech has to be understood within the limitations of the Lancaster House Agreement of 1979, which negotiated the path to the birth of a democratic and independent Zimbabwe after several years of a brutal armed struggle. He did not have much of a choice but to make such a speech in order to initiate the difficult task of

> state-building—and nation building—in an atmosphere of intense suspicion and even hostility on the part of those he had defeated at home; against the background of overt and covert military threats of military, political, and economic destabilization from South Africa.[5]

Thus, one way he sought to do this was to appease the white settlers by underscoring his respect for *nature*.

But even if he may have seemed to have wanted to appease restless whites, Mugabe's commitment to environmental issues remained patently evident throughout the first two decades of his reign. In the early 1980s, he embarked on what Stephen Brain has called "the environmental panacea of choice for dictators in the twentieth century"[6]—namely afforestation. He declared the *National Tree Planting Day* (*NTPD*) as the ideal afforestation approach and made it an annual occasion for the emerging Zimbabwean state. By so doing, Mugabe aimed to inculcate a post-independence sense of what Mahesh Rangarajan has called "ecological patriotism"[7] or, in Carl Death's terms, creating "environmental citizens,"[8] that is, citizens who "take pride in their country and in the fauna and flora indigenous to that country."[9] If Mugabe seemed to falter in his commitment to other environmental issues, especially in the period, 2000 to 2017, the one issue he remained consistently engaged with was the planting of trees and observing NTPD. Even in his old age, at 92 in December 2016, despite struggling to lift a shovel to excavate the soil from the hole, due to ill-health, he still managed to plant a tree.[10] In a clear demonstration of the association between power and his love for nature, in December 2015, Mugabe, accompanied by the First Lady and his two Vice Presidents, broke away from a session of the 15th ZANU-PF annual national people's conference at the exquisite Victoria Falls to plant the symbolic 'tree of the year' to mark the NTPD commemorations.[11] Such was his level of commitment to the environmental program of afforestation and its symbolism that he hardly missed an opportunity to officiate at the annual occasion where he was always honored as the "first citizen" to plant "the tree of the year."

Robert Mugabe, the first prime minister of independent Zimbabwe (1980–1986) and then executive president until he was forced out of power in November 2017, is likely to have an uneven environmental legacy, like other world leaders who sought to make ecological conservation a pillar of state and nation-building, such as Josef Stalin, Benito Mussolini, Jawaharlal

Nehru, Indira Gandhi and Julius Nyerere. While he had his enthusiasts who praised him for being a champion of the environment, he also had his detractors who perceived him as a leader who opportunistically used the environment to consolidate his authoritarian style of governance in the making and consolidation of the Zimbabwean postcolonial state. These seemingly opposing and polarizing views have recently been featured in the Zimbabwean media following his recent fall from grace. A "Climate Story" column in the state-run newspaper, *The Herald*, headlined: "Environment, Climate Thrived under Mugabe,"[12] and heaped praise on the former president:

> One thing is sure: ex-president Robert Mugabe had weaknesses, but under his administration of 37 years, the environment and climate sectors reported marked progress ... From water to wildlife, deforestation to pollution, waste management to energy, the Mugabe presidency enacted one of the most comprehensive legal frameworks for protecting the environment in the world.[13]

On the contrary, *The Herald*'s nemesis, the privately run tabloid, *The Standard* (December 4, 2017), posed the question "Will Mnangagwa Restore the Environment?" and went on to conclude:

> Now that Zimbabwe has a new leadership in the form of President Emmerson Dambudzo Mnangagwa—after 37 long years of Robert Mugabe's rule—there is a lot of hope and optimism in the air. Many have great expectations that, in a bid to capture the hearts and minds of progressive Zimbabweans, the new president will endeavor to right the wrongs and bring back sanity to the country. Mugabe's leadership did not just mastermind the collapse of the economy. Under it, the environment took a beating that will take years to recover from. Under the guise of black empowerment, Mugabe continually gave his cronies the green light to venture into the most environmentally destructive ventures imaginable.[14]

Neither these contested positions nor an assessment of President Mugabe's environmental credentials have as yet received any profound scholarly attention. More importantly, neither position is helpful if abstracted from the historical and political economy context of Zimbabwe and considerations of what influenced his choices and decisions.

Writing about the devotion of Indira Gandhi, former Prime Minister of India (1966–1977 and 1980–1984) to ecological issues, Mahesh Rangarajan delineated the same binary views about Gandhi that have been attributed to Mugabe above.[15] One view extolled her for being an environmental visionary while the other deprecated her as an opportunist whose engagement with nature was only meant to give sustenance to her authoritarian rule.

But Rangarajan posits that these two views are not necessarily mutually exclusive.[16] For him it is as problematic to see Indira Gandhi, or any other leader for that matter, "as intrinsically authoritarian and the other as a benevolent presence assisting nature's recovery"[17] because "one is functional and the latter serves to view political leadership in the abstract, almost in a vacuum, shaping but hardly touched by cultural and intellectual influences of the time."[18] This critical observation about the persona of Indira Gandhi has direct resonances with that of Robert Mugabe only as far as their commitments to environmental issues in the changing historical and political economies of their two different countries are concerned. Even though Mugabe started off with a firm commitment to environmental matters, that fact does not render him a paragon of environmental virtue for all time. Whatever commitments he made were shaped by the political and historical context and challenges as well as opportunities of his time as we shall see below. Similarly, he was far from simply being "intrinsically authoritarian," for that characterization is devoid of context, hence "overly functional" to borrow Rangarajan's phrase, once again.[19]

The Mugabe era can be divided into two broad temporal segments. The first one, from 1980 to 1999, witnessed a more benign, liberal and conciliatory attitude toward the white settler capitalist community and the environment. However, the second one, from 2000 to 2017, was an era in which there was a somewhat leftward/radical slant when Mugabe and his government decisively tinkered with the land question. This chapter documents how Mugabe and his government drifted from a pro-environmental reform philosophy to one where practices to ensure his political survival led to actions which resulted in a diminished commitment to ecological conservation. The chapter demonstrates how international sanctions and the isolation of the ZANU-PF-led government also compelled Mugabe to adopt policies that encouraged environmental destruction in Zimbabwe. This chapter advances three interlocked claims. First, it argues that, as the political fortunes of the ZANU-PF government began to wane from 2000 until 2017, the country witnessed an intensified tendency toward authoritarian rule. The violent Fast-track Land Reform program became the most manifest expression of this tendency. In the process, the state adopted a number of environmentally damaging actions. These ranged from the unbridled plunder of Zimbabwe's wildlife to the pillaging of diamonds by local elites and state-sanctioned Chinese mining companies that barely employed environmentally sound practices. In a country in which democratic space was severely curtailed, the institutions (such as the Ministry of Environment and the Environmental Management Agency), that once rendered environmental safeguards, became relatively inept and could hardly challenge the powerful vested interests involved in these acts. These political elite groups were bent on natural resource extraction to satisfy their personal greed or prop up the tottering ZANU-PF party and government now operating under the grip of Western-sponsored international sanctions.

Second, it argues that the actions of "free democracies" such as Britain and the United States had negative effects on the economy of Zimbabwe that Mugabe found quite difficult to circumvent. This led Mugabe to increasingly up the ante against the West and white capital still in control of aspects of the Zimbabwean economy and began to call for "resource nationalism."[20] Third, I argue that these two factors are very central to understanding Zimbabwe's environmental issues, or any postcolonial state for that matter. It is critical to not lose sight of the actions of very powerful external forces in writing Zimbabwe's environmental history. Britain, Europe's oldest democracy, robbed and raped Zimbabwe of its resources during the colonial days. Later, in the 2000s, "free democracies," that is, Britain and its allies, isolated the country for political reasons, which resulted in further environmental ruin. Without necessarily portraying Mugabe as a victim of international sanctions, for he is not inculpable, Zimbabwe certainly had very little space to navigate politically and economically. The actions taken to ensure political survival, for example the seizure of land, resulted in a far-reaching re-creation of human–nature relations, the effects of which permeated all facets and classes of Zimbabwe. Furthermore, these actions simply undermined the protective environmental laws and institutions that had been founded in the previous period (1980–1999). Mugabe and his government failed to take action against those accused of, or seen to be, violating environmental protocols. Above all, the chapter contends that, even as Mugabe displayed authoritarian tendencies right from the onset of his rule and brooked no opposition, he displayed a pro-environmentalist bias. However, this was to drastically change only when confronted by the clear prospect of losing power from 2000 to 2017. In some ways, this chapter provides a fine example from Africa buttressing Frank Uekötter's proposition that "over the course of the twentieth century, conservation has demonstrated that it can co-exist not only with fascist, but also with socialist and third-world dictatorships of all kinds."[21]

A new beginning: redressing socioeconomic and environmental inequities among the poor, 1980–1999

In the first decade of Zimbabwe's independence (1980–1990), Mugabe's government was determined to redress the social and economic inequities of the colonial past affecting mostly the African majority. This was exemplified by dedicating each of the first four years to themes focused on transferring political power, even if in name only, to the *povo* (the ordinary people) and consolidating that power as well as transforming their socioeconomic status. Thus, 1980 was dubbed "*Gore Remasimba eVanhu*" (Shona) or "*Umnyaka Wamandla aBantu*" (Ndebele), or "Year of the People"; 1981 was named "*Gore Rekutsigira Masimba eVanhu*," or "Year of Consolidating People's Power," 1982 was labeled "*Gore Reshanduko*," or "Year of Transformation," and 1983 was styled "*Gore Rekugutsa Ruzhinji*," or "Year of Meeting People's Needs." It is, however, not clear how much power, in real terms, was devolved to the "povo." However, there are a few areas in which the Zimbabwe government clearly made an indelible mark in terms of changing and improving the lives of the majority of the African population in Zimbabwe, namely, the massive development of education and health services for the public as well as the betterment of ordinary people's access to these.[22] In addition, as far as the environment was concerned, the Zimbabwe government was quick to demonstrate its commitment to the welfare and conservation of nature by devoting itself to arresting processes perceived to have led to environmental degradation, soil erosion, deforestation and the desertification of Zimbabwe's landscapes. There is no doubt that Mugabe's government was alive to the hitherto colonial state's unsound environmental policies. In December 1982, Victoria Chitepo, the new Minister of Natural Resources and Tourism, clearly summed it up:

> The current state of Zimbabwe's natural environment was not healthy because of the disastrous socio-economic policy of the colonial regime, particularly the land policy that concentrated large populations on land ecologically unsuited to carry it. My ministry has been grossly perturbed by the actions of some of our compatriots in the handling of some of our crucial and basic natural resources such as trees, soil and water.[23]

The new government immediately grappled with these inherited environmental problems. Recognizing "the need for an integrated approach to the management of natural resources,"[24] the Mugabe government introduced a range of positive people-oriented changes by making a "paradigmatic shift from state-centered control toward alternative conservation and natural resource approaches in which local people would play a central role."[25] It developed a National Conservation Strategy (NCS) for the express purpose to "draw the attention of Government and the populace to the need for conservation and environmental protection by documenting the state of the environment."[26] This was soon followed by the creation of a new government ministry responsible for the environment to synchronize all environmental and conservation issues that were at stake. Public land reclamation activities, including gully reclamation and reforestation were introduced.[27] Unlike before, it was now a requirement for an Environmental Impact Assessment to be carried out "prior to the construction of any massive infrastructure."[28] Victoria Chitepo, Minister of Natural Resources and Tourism, underscored the importance of making rational, scientific and exhaustive investigations into the socioeconomic, ecological, and environmental implications before embarking on the construction of massive infrastructure such as hydroelectric schemes: "All major development plans had to take into account the need for maintaining the balance of nature and the ecosystems and the preservations of a healthy environment."[29]

In line with this thinking, one of the first programs the government embarked upon to redress the aforementioned ecological imbalances was when Mugabe declared the first Saturday of every month of December as "National Tree Planting Day," making this day "an important part of Zimbabwe's environmental calendar."[30] In subsequent years, and in ritualistic fashion, the same speech that Mugabe delivered on this day would be read on his behalf by cabinet ministers and senior government officials across the country where ceremonies to commemorate this day were being held. National Tree Planting Day was declared to achieve three main objectives: to "motivate the nation to plant and conserve trees; enlighten the nation on the importance of forest and woodland resources; and enhance household food security."[31] An article which appeared in *The Herald*, intended to inform the public, "How the Green Revolution is Taking Root," feted the Prime Minister as a "crusader" of tree planting.[32] Mugabe saw the struggle against deforestation as linked to the transformation of ordinary people's lives and developing a sense of aesthetic beauty. Revealing his intense eye for natural beauty, Mugabe evocatively extolled the virtues of tree planting and creating forests:

> This year the national emphasis is on transformation. Let there be forests throughout our nation. Let every treeless area of the country be replaced with trees. Transform the countryside into a place of beauty and utility. National Tree Planting Day should not just be a ceremonial day. It marks the beginning of the tree-planting season. I would remind you that once trees are planted, they must not be neglected. There is a need to tend them until they mature.[33]

In mobilizing and exhorting the people to "transform the countryside," Mugabe was not only interested in his historical and environmental legacy but was also genuinely responding to what was evidently an ecological crisis in the making. This program has continued to be observed every year in Zimbabwe, albeit, not with the same early independence, pomp-and-ceremony practice, due to the dire macro-economic problems the country has faced.

In the mid-1980s, Mugabe continued with his lofty ideals of redressing the socioeconomic inequities of the past for the ordinary *povo* in the countryside. It was the *povo*, which had essentially given the ruling party its rural base and material succor in most of Zimbabwe from which to launch the armed struggle, which helped to dislodge Smith. It was now time for the Mugabe government to recompense the masses. Recognizing that hunger afflicted the lives of many communal farmers, Mugabe sought to boost the capacity for communal lands' agricultural productivity.

But he was also acutely aware that this could not possibly materialize on ecologically depressed lands. Mugabe comprehended that agricultural growth, restoration, and safeguarding of the environment were inextricably intertwined features. Fighting hunger thus became his other commitment upon assuming the office of Prime Minister of Zimbabwe. The state provided massive agricultural implements, and research extension and marketing services were provided to Black communal farmers. This injection of agricultural capital by Mugabe's government resulted in a phenomenal increase in maize and cotton production in the first-half of the 1980s—which became known as "Zimbabwe's Agricultural Revolution" and earned Zimbabwe the moniker the "breadbasket" of Southern Africa.[34] In recognition of his efforts to end hunger, Mugabe became only the second recipient of the $100,000 Africa Prize for Leadership for the Sustainable End of Hunger in 1988 from the New York-based philanthropy and aid organization, Hunger Project.[35] He was lauded by his hosts for turning Zimbabwe into a self-sufficient food producer leading to portrayals of Zimbabwe as the "agricultural success story" of Africa.[36] If he convinced his audience that agricultural growth and environmental protection were the flip sides of the same coin, he must have gratified them when he observed: "The struggle against hunger and malnutrition is also a function of democracy."[37] He attributed Zimbabwe's accomplishments to a rural resettlement program for landless African peasants initiated in the 1980s.[38] In addition, more African communal farmers—about 100,000—were producing cotton in the early 1980s, all of which was picked by hand thus providing both employment and ensuring good quality, hence attracting buyers on international markets. Communal farmers also began to venture into tobacco farming and guaranteed Zimbabwe's position as one of the leading tobacco exporters in the world. At the same time, in 1986, the government urged farmers to venture into other spheres of farming which saw mostly white commercial farmers investing in horticulture and game, ostrich ranching and crocodile farming.[39]

But the image of a food self-sufficient Zimbabwe began to change especially in the early 1990s as the sparkle on the term "agricultural revolution" started to fade away. The food production revolution in the communal areas was short-lived because of a changing political economy in Zimbabwe such as a slow resolution of the land question since independence, malnutrition among 30 percent of the country's school children, and a sluggish agricultural growth in the late 1980s and early 1990s coinciding with a devastating drought in 1991–1992 and recurrent droughts throughout the 1990s and the 2000s. Clearly, while Mugabe's agro-strategy had yielded positive results, the range of other challenges that the country faced—unemployment and poverty both roots of hunger and malnutrition—were evident. This raised the need for a new approach that went beyond agriculture to invigorate the rural economy. Agricultural and scientific institutions also had to generate new solutions with new technologies as well as the marketing of new crops.[40]

However, new policies had to be adopted in conjunction with environmental degradation. As Rukuni has argued, given the precipitous growth in population and how this led to massive deforestation as well as farming in delicate environments, the consolidation of agriculture in the rural areas had necessarily to be part of the solution to the problem of environmental decline, especially deforestation, soil erosion, and silting.[41]

Adding to the list of challenges confronting Mugabe was the energy crisis and its ecological implications for agricultural and environmental decay. People in rural areas were heavily dependent on wood fuel which they procured from vanishing forests. In short, therefore, the challenge that Mugabe faced in the 1990s onwards was to spawn the requisite political will to make enduring investments in both physical and communication infrastructure to generate fresh economic prospects for the rural areas over the next couple of decades. By the mid-1990s, it was very clear that, notwithstanding the accomplishments, land policy remained one of the biggest headaches for Mugabe as he was still constrained by the provisions of the Lancaster House Agreement of 1979. It still remained one of the foremost political and racial issues affecting the country. A lasting solution to the land question was key to achieving "both efficiency and equity goals" as was arresting environmental deterioration.[42]

Another site of environmental governance, where Mugabe provided the communal farmers with restitution, was involving the rural folk in community-based conservation programs. Assisted by non-governmental organizations (NGOs), the Zimbabwe government established the CAMPFIRE idea. This concept was essentially meant to give primary powers and responsibility to villagers to make decisions pertaining to the management of wildlife and other local natural resources (woodlands, water, and grass) and not to local government institutions. By the end of 1989, over a dozen rural districts had become involved in CAMPFIRE, and by 1995 twenty-six districts were involved. Every district situated near a national park was given authority to be part of the CAMPFIRE program. CAMPFIRE projects were, however, plagued by difficulties such as administrative problems, policy issues and corruption, interference from political parties, clashes between institutions and politicians and chiefs and other traditional leaders. All these environmental projects were being introduced across the nation in a period of relative peace for the first three years of independence. However, this trend was to be interrupted in ways that demonstrate what happens to environmental conservation efforts when a government turns to authoritarian measures to deal with conflict, and how war and insecurity impact on the environment, as transpired during the outbreak of civil war in western Zimbabwe from 1983 until 1987. The years of peace after the 1987 Unity Accord that was struck between the warring parties, ZANU-PF and PF-ZAPU, did not necessarily translate into economic growth for Zimbabwe. New challenges emerged as political repression against opposition forces to Mugabe's government intensified their criticism of its performance. By the end of the 1990s, however, it was very clear that the Mugabe government was losing political support. The end of the 1990s marked the end of a relatively political and environmental period, to be superseded by the beginning of a new politically, economically, and ecologically contentious period in the 2000s until 2017.

Political and economic decline and its environmental impact, 2000–2017

During this period, the Zimbabwean government perceived the newly formed Movement for Democratic Change (MDC) as an enemy of the state supported by Western powers aimed at regime change. In order to forestall the MDC's growing influence, Mugabe unleashed a torrent of authoritarian measures to clip its wings, in order to weaken its appeal to the populace as a formidable alternative to ZANU-PF. From 2000 to 2017, Zimbabwe went through a series of multiple and interwoven crises.[43] Except for the period from 2009 to 2013, when a Government of National Unity (GNU) comprising the two MDC formations[44] and ZANU-PF governed the country, Zimbabwe's economy was in the doldrums.[45] During this short interregnum, the GNU adopted pragmatic policies which earned it confidence from the Zimbabwean populace, foreign countries and monetary institutions. However, this promising economic resurgence was short-lived when the ZANU-PF party chose not to continue with the GNU and pushed for the holding of elections in 2013. It "won" the elections but their credibility was challenged by the MDC and other civil society organizations. After 2013, runaway economic decline continued relentlessly "as the ZANU-PF government seemed not to have any clue about how to arrest the economic meltdown."[46] With the contraction of the economy, the job market shrank and the vast majority of Zimbabweans were forced to turn to the informal sector to eke out an existence. For example, "in 2013, reportedly, 84 percent of the country's employed population of 5.4 million was in the informal employment, while only 11 percent remained in formal employment and the balance was not classified."[47] The period leading up to the GNU (2000–2008) and the period after (2013 to 2017) have been the worst for Zimbabwe since its attainment of independence, as the ruling party, facing political and economic challenges and a real/imagined onslaught from the West, embarked on ruthless measures to cling to power. Some of these authoritarian measures, such as the Fast-Track Land Reform program and the violent (*jambanja*) land invasions and seizure of white-owned commercial farms, the discovery of diamonds in Marange in Manicaland and how the state managed their extraction, urban governance and pollution, directly and indirectly had deleterious consequences for both the rural and urban environments. What then was the relationship between Mugabe's increasingly repressive state and the environment?

The fast-track land redistribution exercise that began in the 2000s was a messy affair. The equitable redistribution of land, had it been done in an orderly way, was a historically justified process. However, the manner in which it was carried out had far-reaching negative consequences for the Zimbabwean economy, the manufacturing sector (it led to de-industrialization), agriculture, and the environment. Since March 2000, thousands of white-owned large-scale commercial farms were violently seized and occupied by landless squatters, many of whom masqueraded as "war veterans" of Zimbabwe's armed struggle. The impact of this violence on the Zimbabwean economy has now been well documented. But little scholarly attention has been paid to the impact of these violent land seizures on the environment. In most cases, the ecological impact of the conflict has been subsumed under its negative effects on agriculture. Although this was not a war situation, the conflict and insecurity engendered by the violence attendant to the land seizures was of such magnitude as to undermine conservation efforts for quite some time.[48] Nowhere is this more evident than in the production of tobacco, which has led to massive deforestation in the country.

The seizure of farms saw many small-scale African farmers enter into tobacco farming given its lucrative appeal if well managed. To give a sense of numbers, a *Herald* reporter stated that "in comparison with the 2012 growing season, the number of tobacco growers registered in the 2013 growing season increased by 22,000 to 64,775."[49] While this opened up opportunities for entrepreneurial growth for African peasants, the downside was that tobacco farming has a destructive effect on vegetation. The same reporter candidly stated:

> This increased production has been happening against the backdrop of massive deforestation in the countryside, which is a cause for concern. Is the country prepared to face the inherent environmental challenges of embracing the mainstreaming of previously disadvantaged people into the tobacco sector?[50]

The reporter was hinting at the state's lack of long-term vision about the implications of a quick, unplanned, un-financed land redistribution exercise that did not take into cognizance the potential for environmental disaster. What the Fast-Track Land Reform program did was not only give the previously landless peasants access to land, but it also opened up access for them to many other resources—for example, indigenous forests from which they had been alienated by colonial laws such as the Land Apportionment Act and which white commercial farmers had preserved for tourism and other purposes. The problem with using indigenous trees for immediate economic benefits at the expense of the environment is that "indigenous trees take a very long time to grow to sizes that can then be used to cure tobacco."[51] These views are supported in a report entitled *Inside Zimbabwe*, from the conservative Mike Campbell Foundation, which also lamented the "environmental catastrophe" that these small-scale farmers were in the process of creating:

> Last year [2014] 350,000 hectares of indigenous timber, mainly *msasa* forests, were cut down, much of it to flue-cure tobacco. The farmers do not have the infrastructure, financial resources or means of transport to transfer the necessary coal from the Hwange collieries in the west of the country as their predecessors [white commercial farmers] did, so they have taken to the most accessible form of fuel in the area—the *msasa* forests. So serious is the problem that foreign diplomatic missions have confronted the government over the issue and one diplomat told me "we are hoping they (the government) have the sense to realize this has to stop. It is terrible. It is people going for the short-term solution."[52]

Although the country's environmental watchdog, the Environmental Management Agency (EMA), established in 2004, has been policing farmers poaching trees for the aim of curing tobacco, it is not clear whether its activities have had any lasting effect at all. It is also unclear how those people who are struggling to make ends meet would be able to pay a statutory fine of US$300 per tree if caught indiscriminately and illegally cutting down trees.[53] If these levels of deforestation are not contained or stopped completely, this may end up undermining the survival of the same people who are trying to salvage themselves from the depths of poverty.

A negative social, economic, and environmental downside of the Fast-Track Land Reform program was not only the destruction of farm infrastructure, such as water irrigation systems (water pumps and pipe), but also the massive displacement of more than 2 million black farm workers and their families, as well as the disruption of educational activities on the farms. This left a massive number of homeless and unemployed within the rural community who obviously found it hard to survive.[54] The once thriving agricultural enterprises, such as the Kondozi Estates (Manicaland province), the Anglo-American Corporation-owned Mazowe Citrus Estates (Mashonaland Central), and the McIntyre Estate (situated between Harare and Norton) were rendered derelict by the land invasions. The case of the Kondozi Estate is perhaps the most controversial one as its takeover allegedly involved Mugabe's government cabinet ministers—who seized the estate from a black Zimbabwean main shareholder (owned 52 percent of shares) who ran it jointly with a white Zimbabwean. A year before Kondozi Estate was invaded by squatters with the alleged involvement of the Zimbabwe National Army in April 2004 at the behest of some cabinet ministers, the estate was described as a "highly successful, striving and profitable agriculture business enterprise which at its peak in 2003 generated US$15 million in profits" and "employed an estimated 10,000 families and other out-growers."[55] Without going into the details of the political saga surrounding the seizure of the Kondozi Estate, suffice to highlight that it was laid to such waste that to this day it has not recovered. A former senior employee reported: "Look at Kondozi today, it is like a desert. We used to live a good life but since ZANU-PF took away the farm, the workers were left suffering and up to now some are still living in abject poverty."[56] Former employees bemoaned how their children had lost access to education and how the social fabric within families had been broken asunder by the disruption of farming enterprises at Kondozi. One had this to say: "Our girl children have ventured into prostitution. There have been family breakdowns, as some wives have deserted their homes and have been married by illegal diamond dealers at nearby Chiadzwa diamond field."[57] The mention of the Chiadzwa diamond fields is a perfect point at which to turn to yet another salutary example of the inextricable linkages between state authoritarianism, mining, and environmental degradation which unfolded during the climax of the Zimbabwean political and economic crisis under Mugabe's watch.

State authoritarianism, diamond mining, pollution, and environmental degradation in Chiadzwa, Marange in Manicaland Province, 2006–2016

Just at the height of Zimbabwe's economic and political meltdown, discussed above, a massive discovery of diamonds was made in September 2006 in a little-known part of Chief Marange's area in the eastern province of Manicaland in Zimbabwe, called Chiadzwa. This discovery came at a time when both the Zimbabwean state and ordinary people were struggling for survival. The local people in Chiadzwa began informal mining activities to obtain diamonds for resale. However, the Zimbabwe government seeing a possible panacea to its economic woes, wasted no time in ensuring that Chiadzwa was under heavy state control and let loose both the police and army on local people as from 2008. To date, "the Marange area … is under tight security control … only authorized vehicles and people are allowed through the checkpoints into the Marange area."[58] Such were the conditions under which villagers in the Marange area lived. The government also instructed its parastatal organization, the Zimbabwe Mining Development Corporation (ZMDC) to enter into joint venture deals to process the diamonds in Chiadzwa.[59] The ZMDC went into several partnerships with companies that had interests in diamond mining from South Africa, Canada, Dubai, and China. Thus, large-scale mining and processing of diamonds began in earnest in 2009 and was dominated by "foreign-owned companies in joint ventures with the elite-run state-owned ZMDC."[60]

No sooner had mining started than local people in particular began to experience numerous environmentally hazardous problems and general damage to their environmental landscapes. As Duri has noted,

> Mining operations involving heavy earth-moving equipment and the dumping of mine effluent caused massive siltation of the Save and Odzi Rivers and threatened the livelihoods of people and livestock in the districts of Chipinge, Chimanimani, Mutare and Buhera. Commercial

mining activities destroyed the landscape in Chiadzwa and rendered the land virtually useless.[61]

Local communities and NGOs such as the Zimbabwe Environmental Lawyers Association (ZELA) protested the harmful effects posed by diamond mining firms at Chiadzwa. Paradoxically, senior Zimbabwean politicians and other senior government officials in Manicaland province were also critical of the activities of these firms—which Duri has characterized as simply wanting "to gain political mileage among the local inhabitants given the fact that the ZMDC, a state-owned company was involved in joint mining ventures with foreign companies and was partly responsible for the environmental catastrophe."[62] Marange communities also condemned commercial diamond mining companies for pursuing destructive practices on fauna and flora thereby undermining their livelihoods. Perhaps the most censured diamond mining company was the Chinese company, Anjin, formed as a joint venture also involving a Zimbabwean company and a Chinese firm. Anjin was accused by Marange villagers of desecrating the graves of their departed loved ones as they mined for diamonds. Their demands for compensation from Anjin were unheeded and when they sought political intervention after approaching their own chief, the Manicaland Provincial Administrator hit back at the villagers and censured them for not following the proper channels: "The community was supposed to have directed the problem to us and now without our involvement, it is difficult to intervene."[63]

But it is not just environmental abuses that have been the hallmark of these mining companies. A recent *Mail & Guardian* newspaper article entitled "Zimbabwe's Diamonds Bleed People," reported that "mining by the state has brought human rights and environmental abuses to Marange villagers."[64] Notably, international and local NGOs such as Human Rights Watch, Global Witness and ZELA have reported abuses perpetrated by private security companies in the employ of the mining companies in Marange and also "the Zimbabwean Army, police and Central Intelligence Organization."[65] Communities were also removed from Chiadzwa without any compensation despite the existence of constitutional guarantees protecting the rights of ordinary people faced with such removals. While on a fact-finding investigation, members of a parliamentary portfolio committee on mines and energy as well as ZELA witnessed cases of abject poverty and deprivation at one of the areas, Arda Transau', where villagers had been relocated to.[66] The government's pursuit of profit at the expense of local peoples' human rights and the environment stand in marked contrast to its early 1980s commitment to ensuring that the natural environment was rehabilitated and made to benefit the poor. However, the authoritarian policies of the Zimbabwe government did not spare the urban areas from environmental damage that has, over time, left the majority of urban dwellers in most, if not all, Zimbabwe's cities vulnerable.[67]

Infrastructure development on urban wetlands at the expense of the environment

As a result of the multiplicity of crises hinted at above, even the urban ecosystem—particularly biodiversity sites such as wetlands—came under severe threat after 2000 as a result of a number of socioeconomic and political factors. The invasion of wetlands in urban areas, especially in metropolitan Harare, has become rampant over the last two decades, leading one conscientious newspaper reporter to lament in his column "Climate Story" that "the destruction of wetland areas in the country is unsustainable. No amount of environmental legislation or public outcry has so far managed to effectively stop wetlands from turning into drylands or built-up lands. The damage is severe."[68] Reflecting the ongoing power struggles between the ZANU-PF central government and the MDC-run City of Harare, it has constantly been alleged that bigwig politicians in the ruling party and even at government level have authorized the building of infrastructure and houses on wetlands by overriding the municipality which has the legal authority to protect such wetlands as it has done ever since it was run by ZANU-PF councilors. Perhaps the most well-known case was when the ZANU-PF government—to sweeten deals it had entered into with the Chinese government as part of its "Look East Policy" when the West had shunned it for various economic and political reasons—sanctioned the construction of a massive 62,000 square meter Chinese hotel and shopping mall called Long Cheng Plaza, along the Harare-Bulawayo highway road and near the National Sports Stadium, on a well-known wetland. The following quote captures the sense of despair of one concerned citizen, Jeffery Gogo, who was disgusted by this construction: "Where peaceful plush green indigenous trees once stood, now stands two large fierce lion-like creatures, as you enter the gates of this sprawling shopping and hotel complex. Here, environmental considerations were brutally trashed by desires for infrastructure development."[69] Indeed, in the mid-1980s indigenous trees had been planted on the wetland by diplomats accredited to Zimbabwe and a beautiful "forest" had emerged over time. But this was allowed to disappear to make way for this development.[70] A comment to the editor of the privately owned *Daily News* was blatant in its condemnation of the disregard for the environmental laws of the city:

> this is a shame. On one hand (President) Mugabe says we want indigenization ... and on the other allows the Chinese to ruin our wetlands and destroy the environment in the process. Is there any Zimbabwean mall in China? It is ridiculous that someone actually approved such a project despite the obvious problems and short cuts.[71]

But it is not only the influential Chinese who were being given access to wetland areas. Ordinary people were also being allowed to build their houses on these fragile ecosystems without properly applying for land approved for human habitation. Invasion of wetlands has taken place in suburbs such as Hatcliffe, Borrowdale, Dzivarasekwa Extension, Kuwadzana Extension, and other previously "sacred" spaces such as the KuMbudzi Cemetery in the eastern side of Harare. It appears that the invasion of these spaces such as wetlands has even increased as the ZANU-PF government is distributing residential stands, particularly to youths. The talk of the town was that the reason why a number of informal settlements—several of these positioned on wetlands and river banks have burgeoned—was that ZANU-PF was "dangling land as a carrot to woo urban voters" away from the MDC, particularly leading up to the 2013 elections. Hence political power today is being maintained to the detriment of the environment. The environmental consequences of building on wetlands became apparent during the 2016/2017 rain season that resulted in flash floods that exposed the vulnerability of newly resettled residents, most of whom benefited from ZANU-PF urban residential stands programs.

Conclusion

Upon assuming power in 1980, Robert Gabriel Mugabe made a magnanimous gesture toward his erstwhile enemies to the surprise of many people who had thought that he would take revenge on those against whom he had waged the armed struggle. While this is almost common knowledge now, a rather less known fact is how he used an environmental conservation congress speech to underscore his commitment to national reconciliation. But what that speech is not acknowledged for are its ecological overtones and how he vowed that "the laws of nature and conservation" had to be respected. He lived up to this commitment and engaged with environmental issues making them part and parcel of nation and state-building of the new country. Despite creeping authoritarian inclinations right from the onset of his rule and intolerance for political opposition, he displayed a pro-environmentalist bias especially in the first two decades of his rule. His declaration of the National Tree Planting Day as an annual event became the symbol of his engagement with nature. He understood the nexus between poverty especially among Africans living in communal areas of the country and environmental degradation where it was rife. But he was also very clear about the connections between stimulating economic growth and arresting land degradation as the only viable features for ending hunger and poverty. These are the principles and ideals that seem to have guided his agricultural and environmental policies, especially in the first two decades of independence as he led Zimbabwe to become an agricultural success story in the 1980s for which he received a prestigious Hunger Prize. They were born of critical assessments of what was wrong on the ground and among the rural peasants in particular. Although there were threats of political destabilization because of the "dissident" problem leading to the infamous

Gukurahundi massacres of people in the province of Matabeleland, and growing opposition from other smaller parties, he remained unperturbed by these issues until such time that his power was demonstrably threatened toward the end of 1999. The brutal suppression of the "dissident" problem in the early to mid-1980s was already earning Mugabe the descriptor "authoritarian," but at this point in time there were no rising environmentalist movements that posed enough of a threat to him to have warranted censure from him or his government. His engagement with environmental issues in the first two decades of independence should not cast him as a benign ruler because he had his own failings. When the celebration around the post-independence "agricultural revolution" attributed to his pragmatic policies of supporting smallholder communal farmers began to wane, it was quite clear that this was a momentary success. In short, this agricultural revolution did not bring to an end hunger nor did his environmental policies as such.

As the decade of the 1990s drew to a close, the ensuing period, 2000–2017, presented a remarkable political shift for President Mugabe. This period was his most challenging for all time as the leader of postcolonial Zimbabwe as many looming conflicts and tensions that had been fermenting since the 1980s and 1990s finally unraveled in the early 2000s. Intense opposition to his rule came to a head when his party almost lost the 2000 parliamentary and mayoral elections to the MDC and when he also nearly lost the presidential elections in 2001 and 2008. Realizing this risk to his political survival he went into a tailspin and initiated a massive, unbridled, violent Fast-Track Land Reform program which spread across the length and breadth of the country and resulted in the ejection of white commercial farmers from their lands. For all his authoritarian proclivities Mugabe believed and, indeed, sold this land reform exercise as an altruistic move meant to benefit the African people whose lands were seized from them by white settlers from the 1890s until independence in 1980. This represented, at least in Mugabe's scheme of things, an attempt to finally reorder in a politically grounded way, society–nature relations that had remained racially skewed since the days of colonialism. He felt compelled to use autocratic means to finally deliver on the land question, which he had failed to do for nearly three decades but only for purposes of political expediency as land redistribution seemed to be the only instrument left for him to use to cling to power. Mugabe's state-sanctioned and unbridled land seizures set a train of environmental destruction and led to the reordering of society–nature relations which benefited, in large measure, the well-connected political elites. As he was vehemently criticized for his policies and targeted sanctions imposed on him and senior members of his party and government by Western countries, he dug in and resorted to resource nationalism as a way for his country to survive economically. In all these years of turmoil in Zimbabwe, Mugabe neither said anything nor took any action against those big violators of environmental policies, as he would have done in the past. His land reform exercise and the problems it engendered led to the emasculation of the formidable body of environmental laws and institutions he had built up over the three decades he was in power. Mugabe started off as a benevolent and ecologically minded statesman who remained committed to improving human–nature dynamics in his country for his people for as long as he remained comfortably ensconced in his seat of power. However, as his stay in power was challenged he took recourse to authoritarian measures that impacted on the environment directly or indirectly. Now out of power, Mugabe leaves a mixed environmental legacy. Since his rule, President Mnangagwa has reversed some of the policies that had been negatively impacting the environment and there are initial indications that his administration is determined to reinstate the rule of law in many spheres of the political economy as well as the environment of Zimbabwe.

Notes

1 This chapter was first presented at the *Commanding the Environment or Green Dictatorships? Nature-Culture-Nature-Society Relationships in Authoritarian Regimes*, International Seminar at the Aleksanteri Institute, University of Helsinki, Finland, April 27–28, 2017. It was originally titled: *The Political Economy of Zimbabwe's Environmental Decline in a Time of Crisis, 2000–2017*.
2 See "The Text of an Address Given at the Opening of the First National Conservation Congress of the Natural Resources Board of Zimbabwe by the Prime Minister, R. G. Mugabe, on 16 September 1980," *The Zimbabwe Science News* 14, no. 2 (December 1980): 287–289.
3 I. Mandaza, "The State in Post-White Settler Colonial Situation," in I. Mandaza (ed.), *Zimbabwe: The Political Economy of Transition, 1980–1986* (Dakar: Codesria Book Series, 1986), 41–42.
4 M. Musemwa, "*Sic utere tuo ut alienam non laedas*. From Wanton Destruction of Timber Forests to Environmentalism: The Rise of Colonial Environmental and 'Sustainability' Practices in Colonial Zimbabwe, 1939–1961," *Environment and History* 22, no. 4: 521–559.
5 Mandaza, "The State in Post-White Settler Colonial Situation," 42.
6 S. Brain, "The Great Stalin Plan for the Transformation of Nature," *Environmental History* 15 (2010): 670.
7 M. Rangarajan, "Striving for a Balance: Nature, Power, Science and India's Indira Gandhi, 1917–1984," *Conservation and Society* 7, no. 4 (2009): 299.
8 C. Death, *The Green State in Africa* (New Haven, CT: Yale University Press, 2016), 139.
9 Death, *The Green State in Africa*: 139.
10 *News24*, 2016/12/20. Accessed: June 28, 2018.
11 "Mugabe Plants Tree of the Year," https://bulawayo24.com/index-id-news-sc-local-byo-79118.html. Accessed: June 28, 2018.
12 "Environment, Climate Thrived under Mugabe," www.herald.co.zw/environment-climate-thrived-under-mugabe-but-only/. Accessed: June 14, 2018.
13 "Environment, Climate Thrived under Mugabe," www.herald.co.zw/environment-climate-thrived-under-mugabe-but-only/. Accessed: June 14, 2018.
14 "Will Mnangagwa Restore Environment?" www.thestandard.co.zw/2017/12/04/will-mnangagwa-restore-environment/. Accessed: June 14, 2018.

15 Rangarajan, 'Striving for a Balance': 299.
16 Rangarajan, 'Striving for a Balance': 299.
17 Rangarajan, 'Striving for a Balance': 299.
18 Rangarajan, 'Striving for a Balance': 299.
19 Rangarajan, 'Striving for a Balance': 299.
20 Defined here as a penchant of governments and people to exert control over natural resources within their countries, especially where such control was dominated by foreign capital.
21 F. Uekötter, "Green Nazis? Reassessing the Environmental History of Nazi Germany," *German Studies Review* 30, no. 2 (2007): 268.
22 A. S. Mlambo, "From an Industrial Powerhouse to a Nation of Vendors: Over Two Decades of Economic Decline and Deindustrialization in Zimbabwe, 1990–2015," *Journal of Developing Societies* 33, no. 1 (2017): 1–27.
23 "The Roots of our Future—Chitepo," *The Herald*, December 4, 1982.
24 CBPR Database—Zimbabwe, Centre for International Environmental Law, www.ciel.org/Publications/CBPR_Zimbabwe_9-20-06.pdf. Accessed: April 14, 2017.
25 CBPR Database—Zimbabwe, 2006.
26 CBPR Database—Zimbabwe, 2006.
27 CBPR Database—Zimbabwe, 2006.
28 CBPR Database—Zimbabwe, 2006.
29 "Chitepo Handed Rivers Report," *The Herald*, November 30, 1982.
30 www.ema.co.zw/index.php/86-national-tree-planting-day.html. Accessed: April 14, 2017.
31 www.ema.co.zw/index.php/86-national-tree-planting-day.html. Accessed: April 14, 2017.
32 "Mugabe Spearheads Tree Planting Crusade: How the Green Revolution is Taking Root," *The Herald*, December 4, 1981.
33 "Mugabe spearheads Tree Planting Crusade: How the Green Revolution is Taking Root," *The Herald*, December 4, 1981.
34 M. Rukuni, "The Prime Movers of Zimbabwe's Agricultural Revolution," in M. Rukuni and C. K. Eicher (eds.), *Zimbabwe's Agricultural Revolution* (Harare: University of Zimbabwe Publications, 1994), 2–3.
35 L. Cliffe, "The Conservation Issue in Zimbabwe," *Review of African Political Economy* (1988), no. 42, 48.
36 "Mugabe Wins Anti-Hunger Prize," www.nytimes.com/1988/09/16/world/mugabe-wins-anti-hunger-prize.html. Accessed: June 25, 2018.
37 "Mugabe Wins Anti-Hunger Prize," www.nytimes.com/1988/09/16/world/mugabe-wins-anti-hunger-prize.html. Accessed: June 25, 2018.
38 "Mugabe Wins Anti-Hunger Prize," www.nytimes.com/1988/09/16/world/mugabe-wins-anti-hunger-prize.html. Accessed: June 25, 2018.
39 Rukuni, "The Prime Movers of Zimbabwe's Agricultural Revolution," 2.
40 Rukuni, "The Prime Movers of Zimbabwe's Agricultural Revolution," 2.
41 M. Rukuni, "The Evolution of Agricultural Policy, 1890–1990," in M. Rukuni and C. K. Eicher (eds.), *Zimbabwe's Agricultural Revolution* (Harare: University of Zimbabwe Publications, 1994), 32.
42 Rukuni, 'The Evolution of Agricultural Policy, 35.
43 S. Chiumbu and M. Musemwa, "Introduction: Perspectives of the Zimbabwean Crises," in S. Chiumbu and M. Musemwa (eds.), *Crisis! What Crisis? The Multiple Dimensions of the Zimbabwean Crises* (Cape Town: HSRC Press, 2012), ix–xxiii.
44 MDC-T, led by its founder, Morgan Tsvangirayi and a splinter one by Welshman Ncube.
45 Mlambo, "From an Industrial Powerhouse to a Nation of Vendors," 106–110.
46 Mlambo, "From an Industrial Powerhouse to a Nation of Vendors," 107.

47 Mlambo, "From an Industrial Powerhouse to a Nation of Vendors," 107.
48 Zimbabwe and the United Nations Framework Convention on Climate Change: Zimbabwe National Environmental Policy. http://unfccc.int/resource/ccsites/zimbab/legislat/policy.htm Accessed: April 23, 2017.
49 www.herald.co.zw/tobacco-production-and-the-environment/ (November 11, 2014). Accessed: April 23, 2017.
50 www.herald.co.zw/tobacco-production-and-the-environment/ (November 11, 2014). Accessed: April 23, 2017.
51 www.herald.co.zw/tobacco-production-and-the-environment/ (November 11, 2014). Accessed: April 23, 2017.
52 *Inside Zimbabwe* (Mike Campbell Foundation: Restoring Justice-Restoring People, July 16, 2017, 3.
53 www.herald.co.zw/tobacco-production-and-the-environment/ (November 11, 2014). Accessed: April 23, 2017.
54 *Inside Zimbabwe* (Mike Campbell Foundation: Restoring Justice-Restoring People, July 16, 2017, 3.
55 www.radiovop.com/index.php/national-news/14402-kondozi-ghost-haunts-zanu-pf-as-former-co-owner-bares-soul-on-se.html (August 22, 2016). Accessed: April 23, 2017.
56 www.thezimbabwean.co/news/zimbabwe/71303/kondozi-farm-10-years-on.html (April 22, 2014). Accessed: April 23, 2017.
57 www.thezimbabwean.co/news/zimbabwe/71303/kondozi-farm-10-years-on.html (April 22, 2014). Accessed: April 23, 2017.
58 *Mail & Guardian*, April 7 to 12, 2017, 20.
59 Duri, F. P. T. "Environmental Activism from Below: The Case of the Subaltern Against Commercial Diamond-Mining Companies in the Chiadzwa Area of Zimbabwe, 2009–2013," in M. Mawere (ed.), *Development Perspectives from the South: Troubling the Metrics of (Under) Development in Africa* (Bamenda: Bamenda Langaa Research Group and Publishing Common Initiative Group, 2016), 415–434.
60 Duri, "Environmental Activism from Below."
61 Duri, "Environmental Activism from Below."
62 Duri, "Environmental Activism from Below."
63 *Mail & Guardian*, April 7 to 12, 2017, 20.
64 *Mail & Guardian*, April 7 to 12, 2017, 20.
65 *Mail & Guardian*, April 7 to 12, 2017, 20.
66 *Mail & Guardian*, April 7 to 12, 2017, 20.
67 M. Musemwa, "Early Struggles Over Water: From Private to Public Water Utility in the City of Bulawayo, Zimbabwe, 1894–1924," *Journal of Southern African Studies* 14, no. 4 (2008): 881–898; M. Musemwa, "From 'Sunshine City' to a Landscape of Disaster: The Politics of Water, Sanitation and Disease in Harare, Zimbabwe, 1980–2009," *Journal of Developing Societies* 26, no. 2 (2010): 165–206; S. Chigudu, "Health Security and the International Politics of Zimbabwe's Cholera Outbreak, 2008–2009," *Global Health Governance* 10, no. 3 (2016): 41–53.
68 www.herald.co.zw/ema-sleeping-on-the-job/ (February 3, 2013). Accessed: April 22, 2017.
69 www.herald.co.zw/ema-sleeping-on-the-job/ (February 3, 2013). Accessed: April 22, 2017.
70 *News Day*, February 18, 2014.
71 www.dailynews.co.zw/articles/2013/12/17/first-china-mall-opens-in-harare (December 17, 2013). Accessed: April 22, 2017.

Bibliography

Alexander, J. and McGregor, J. (2000). "Wildlife and politics: CAMPFIRE in Zimbabwe," *Development and Change* 31: 605–627.

Beeson, M. (2010). "The coming of environmental authoritarianism," *Environmental Politics* 19, no. 2: 276–294.

Beinart, W. and McGregor, J. (2003). "Introduction," in W. Beinart and J. McGregor (eds.), *Social History and African Environments* (Oxford: James Currey), 1–24.

Bond, P. and Ruiters, G. (2017). "Uneven development and scale politics in Southern Africa: What we learn from Neil Smith," *Antipode* 49, no. S1: 171–189.

Brain, S. (2010). "The great Stalin plan for the transformation of nature," *Environmental History* 15: 670–700.

Chakawa, J. and Nyawo-Shava, V. Z. (2014). "Guerrilla warfare and the environment in Southern Africa: Impediments faced by ZIPRA and Umkhonto *weSizwe*," *Oral History Journal of South Africa* 2, no. 2: 36–47.

Chigudu, S. (2016). "Health security and the international politics of Zimbabwe's cholera outbreak, 2008–2009," *Global Health Governance* 10, no. 3: 41–53.

Chimonyo, G. R., Mungure, S. and Scott, P. D. (no date). "The social, economic and environmental implications of diamond mining in Chiadzwa," unpublished report compiled by the Centre for Research and Development in conjunction with Open Society Initiative and Southern Africa Resource Watch.

Chiumbu, S. and Musemwa, M. (eds.) (2012). *Crisis! What Crisis? The Multiple Dimensions of the Zimbabwean Crises* (Cape Town: HSRC Press).

Cliffe, L. (1988). "The conservation issue in Zimbabwe," *Review of African Political Economy* no. 42: 48–58.

Congleton, R. D. (1992). "Political institutions and pollution control," *The Review of Economics and Statistics* 74, no. 3: 412–421.

Death. C. (2016). *The Green State in Africa* (New Haven, CT: Yale University Press).

Drinkwater, M. (1989). "Technical development and peasant impoverishment: Land use policy in Zimbabwe's Midland Province," *Journal of Southern African Studies* 15, no. 2: 287–305.

Duri, F. P. T. (2016). "Environmental activism from below: The case of the subaltern against commercial diamond-mining companies in the Chiadzwa area of Zimbabwe, 2009–2013," in M. Mawere (ed.), *Development Perspectives from the South: Troubling the Metrics of (Under) Development in Africa* (Bamenda: Bamenda Langaa Research Group and Publishing Common Initiative Group), 415–434.

Gambara, C. Z. (2007). "Environmental impact assessment report for the Marange Chiadzwa diamond project." Institute of Mining Research, University of Zimbabwe, Report No. 165: 1–50.

Gibson, C. C. (1999). *Politicians and Poachers: The Political Economy of Wildlife Policy in Africa* (Cambridge: Cambridge University Press).

Hajer, M. A. (2002). *The Politics of Environmental Discourse: Ecological Modernization and the Policy Process* (Oxford: Clarendon Press).

Hill, R. and Katerere, Y. (2002). *Colonialism and Inequity in Zimbabwe: Conserving the Peace, Resources, Livelihoods and Security* www.iisd.org/pdf/2002/envsec_conserving_5.pdf. Accessed: April 15, 2017.

Keeley, J. and Scoones, I. (2000). "Environmental policymaking in Zimbabwe: Discourse, science and politics," Institute of Development Studies Working Paper 116: 1–39.

Laakkonen, S., Pál, V., and Tucker, R. (2016). "The Cold War and environmental history: Complementary fields," *Cold War History* 16, no. 4: 377–394.

Mandaza, I. (1986). "The state in a post-white settler colonial situation," in I. Mandaza (ed.), *Zimbabwe: The Political Economy of Transition, 1980–1986* (Dakar: Codesria), 21–74.

Manjengwa, J. (no date). "Environment and development issues in Zimbabwe, since 2000 and beyond." Centre for Applied Social Sciences, University of Zimbabwe.

Maravanyika, S. (2013). "Soil conservation and the white agrarian environment in colonial Zimbabwe, 1908–1980." (Unpublished PhD thesis, University of Pretoria).

Marongwe, N. (2000). *A Case of Land Occupations in Zimbabwe* (Harare: Zero Publications).

Martin, D. and Johnson, P. (1981). *The Struggle for Zimbabwe: The Chimurenga War* (Johannesburg: Ravan Press).

Mawere, M. (2013). "A critical review of environmental conservation in Zimbabwe," *Africa Spectrum* 48, no. 2: 85–97.

Mazambani, I. and Mashingaidze, T. M. (2014). "The creation of protected villages in Southern Rhodesia: Colonial mythologies and the official mind, 1972–1980," *The Dyke* 8, no. 3: 72–92.

Mlambo, A. S. (1997). *The Economic Structural Adjustment Programme: The Case of Zimbabwe, 1900–1995* (Harare: University of Zimbabwe Publications).

Mlambo, A. S. (2017). "From an industrial powerhouse to a nation of vendors: Over two decades of economic decline and deindustrialization in Zimbabwe, 1990–2015," *Journal of Developing Societies* 33, no. 1: 1–27.

Musemwa, M. (2008). "Early struggles over water: From private to public water utility in the City of Bulawayo, Zimbabwe, 1894–1924," *Journal of Southern African Studies* 14, no. 4: 881–898.

Musemwa, M. (2009). "Contestation over resources: The miner–farmer dispute in colonial Zimbabwe, 1903–1939," *Environment and History* 15: 79–109.

Musemwa, M. (2010). "From 'Sunshine City' to a landscape of disaster: The politics of water, sanitation and disease in Harare, Zimbabwe, 1980–2009," *Journal of Developing Societies* 26, no. 2: 165–206.

Musemwa, M. (2016). "*Sic utere tuo ut alienam non laedas*. From wanton destruction of timber forests to environmentalism: The rise of colonial environmental and 'sustainability' practices in colonial Zimbabwe, 1939–1961," *Environment and History* 22, no. 4: 521–559.

Mutasa, M. and Ndebele-Murisa, M. (2015). "Biodiversity and human development in Zimbabwe," in T. Murisa and T. Chikweche (eds.), *Beyond the Crises: Zimbabwe's Prospects for Transformation* (Harare: Weaver Press), 199–241.

Nickerson, B. J. (1994). "The environmental laws of Zimbabwe: A unique approach to management of the environment," *Boston College Third World Law Journal* 14, Issue 2: 189–230.

Rangarajan, M. (2009). "Striving for a balance: Nature, power, science and India's Indira Gandhi, 1917–1984," *Conservation and Society* 7, no. 4: 299–312.

Rangarajan, M. (2017). "Nature's nation," www.india-seminar.com/2017/696/696_mahesh_rangarajan.htm. Accessed: June 23, 2018.

Rukuni, M. (1994). "The prime movers of Zimbabwe's agricultural revolution," in M. Rukuni and C. K. Eicher (eds.), *Zimbabwe's Agricultural Revolution* (Harare: University of Zimbabwe Publications, 1994), 1–10.

Sachikonye, L. (2003). 'The situation of commercial farm workers after land reform in Zimbabwe: A report prepared for the Farm Community Trust of Zimbabwe, May 2003." http://archive.kubatana.net/docs/landr/fctz_farm_workers_0305.pdf. Accessed: April 20, 2017.

Shapiro, J. (2001). *Mao's War Against Nature: Politics and the Environment in Revolutionary China* (New York: Cambridge University Press).

Shumba, J. M. (2016). "Zimbabwe's predatory state: Party, military and business complex." (Unpublished PhD thesis, University of the Witwatersrand).

Singh, J. and van Houtom, H. (2002). "Post-colonial nature conservation in Southern Africa: Same emperors, new clothes?" *Geojournal* 58: 253–263.

Uekötter, F. (2007). "Green Nazis? Reassessing the environmental history of Nazi Germany," *German Studies Review* 30, no. 2: 267–287.

Zamponi, M. (2008). "Disciplining nature, land, environment and political coercion in colonial Southern Africa." Working paper no. 1, Centro di Studi Storici E Politici Africa E Medio Oriente, Universita di Bologna.

12 Conclusions and perspectives

Richard Tucker

This volume has been assembled against the background of a prevailing set of ideological presumptions in the Northwestern world, that authoritarian regimes are more ruthless and less accountable than open or democratic governments, therefore causing greater environmental damage with less possibility for public access to information and thus environmental remediation. Those regimes are presumed to centralize power and suppress public debate, serving the strategic priorities of power elites and their financial interests. There is little or no space for civil society to function politically, expressing any sort of public will. Moreover, in contrast to open systems, as a generalization they lack clear legal mechanisms for safeguarding the health of the environment.

In striking contrast are arguments for the environmental virtue of authoritarian efficiency in defining and implementing environmental policies. The authors in this book explore these implications in a considerable range of authoritarian settings. The authors address authoritarian regimes of both left and right, from Soviet republics in eastern Europe to their conservative, military-controlled opposites in Latin America. Most of them are chronologically located during the Cold War era, but in some cases the narratives extend through the transition to independent national regimes in eastern Europe, and contemporary trends in Poland and a postcolonial regime in Zimbabwe. The essays' discussions provide insight into the often-obscure meaning of efficiency, and what evidence there is for distinctive environmental effectiveness in the state systems they analyze.

Taken together, these studies illustrate not a polarity between open democratic and closed authoritarian systems in their environmental significance, but a spectrum of state formations, from open to closed, democratic to totalitarian, egalitarian to hierarchical. Moreover, each government in question evolved in the years of the Cold War and after, in response to changing strategic, economic, and environmental realities.

Rapid acceleration of heavy industry, large-scale agriculture and their energy sources dominated international development in these years. What mattered most for the emergence of environmentally sustainable science, public policy, and public participation was how rapidly environmental agendas were allowed to develop in each country.

In various ways every government under review positioned environmental values within the shared determination to achieve rapid economic development through intensive industrial expansion, the priority that drove both Moscow and Washington in their international interventions. All these regimes shared the ideological priority of urgent technological development at the expense of ecological health, whatever their subsidiary environmental values might be. A major aspect was the heavy pressure to appropriate resources for military industry, as a central driving force of development in the era's environmental impacts. In this book that dimension is particularly evident in the essay on Slovakia, though strategic imperatives underlay developmental pressures in other countries as well.

Three chapters on eastern European countries shed light on the ways Cold War development competition between Communist and capitalist systems shaped environmental priorities and their environmental consequences. The East–West polarity played out rather differently and somewhat more obliquely among the dictatorships of Latin America in the same years. There the driving force for "national-developmentalism," industrialization with major participation by state agencies for energy, mining, and other massive infrastructure projects, was often paired with international capitalist corporations.

In every case, highly polluting industrial development took precedence, and environmental values were secondary or marginalized, with little space for public discontent or organizing until the late 1980s, though this varied significantly from one Soviet or Latin American regime to another. In several countries, limited financial resources at the disposal of the regime played a significant role in that overall trend, but always within the constraints of ideological priorities and intra-governmental elite groups' interests.

Power and wealth generally gravitated to joint military–civilian elites, frequently accompanied by rampant corruption. The personal wealth obtained by powerful people—insiders in autocracies but also corporate executives in putatively democratic systems—has repeatedly been achieved by ruthless extraction of natural resources, constituting a major factor in environmentally damaging operations. Is this factor more prevalent in authoritarian regimes that are not structurally responsible to their citizens? There is some evidence for that in these essays, though the question cannot be carefully weighed without equally detailed discussion of a range of western European and North American democracies.

Eastern Europe under the Soviet system

In the Soviet republics the shadow of directives from the center is visible in these studies as a dimension of Moscow's overarching role in shaping national policies. Stalin's 1948 Great Plan for the Transformation of Nature gave ideological legitimacy to some environmental efforts in the republics, expressing the presumption that the anti-capitalist foundations of Marxism guaranteed ideological harmony in the balance between development and environment.[1]

Two environmentally significant sectors were especially visible: forestry policy and industrial energy policy. Reforestation was a common strategy under various regimes (democratic and authoritarian) as a mark of their environmental concern. However, reforestation often came with the thinly disguised dispossession of rural populations or ethnic minorities in the name of protecting the natural environment. This pattern had a longer history, as exemplified in forestry practice in nineteenth-century Prussia. During the Cold War era, Mao's China was a variant on authoritarian reforestation centrally controlled.

The energy sector was fundamental to economic development, and the environment inevitably experienced long-term consequences. The availability of each alternative for the Soviet republics was generally managed by Moscow. Highly polluting soft coal was primary to industrial development in Czechoslovakia and the German Democratic Republic (East Germany). Petroleum was also vital for eastern Europe's industrialization, though oil is not discussed in detail in these papers. Hydropower was the third, with high environmental stakes in eastern Europe as on every continent. Its impacts are discussed in detail in this book.

In the evolution of official agencies and environmental policies, the position of scientific elites was a key to the location and leverage of environmental concerns. In Russian and East European political cultures there was traditionally a high regard for scientific elites in government circles. How this played out in particular socialist republics is illustrated in the two chapters on Czechoslovakia, where technocratic environmentalism from the 1950s onward concentrated on rapid industrialization and agricultural intensification based on hydropower resources. The prestigious Czech Academy of Science played an important role in public policy formation. In the scientific community in early Cold War Czechoslovakia technocratic developmentalist engineers and agronomists were far more influential than biologists, but the biologists constituted a primary source of proto-environmentalism in otherwise largely closed systems.

Two strikingly different variants of the political process evolved, between the Czech and Slovak regions. In the eastern reaches of the Czech region a sophisticated era-long debate played out between the imperatives of rapid increase in industrial and agricultural production and well-established nature conservation and ecology. Controversy centered in the semi-arid flood plain of the Thaya River in Moravia, which central government interests in the 1950s, reflecting Moscow's priorities, were determined to reshape along the lines of international river basin projects. Subsequently, resisting ongoing controversy, construction of three dams, for flood control and irrigation, created three shallow lakes, with attendant ecological costs. By the 1980s and early 1990s the project was opposed by an alliance of local people and environmentalists, expressing newly mature biocentric ecology science and linked to values of liberal democracy. It is evident now that this was a precursor of the recent trend in Poland.

Throughout the entire 1948–1989 period, Slovakia, under centralized administration in Prague, rather than its own regional capital of Bratislava, had little leverage to challenge the dictates of heavy industrialization, the priority enforced from Moscow and Prague, especially aluminum for military needs. This required construction of hydropower dams to supply the concentrated power that processing aluminum required. A massive aluminum plant, an oil refinery and other large-scale industrial plants appeared, oblivious to environmental consequences. Governmental censorship suppressed information on environmental degradation, so it was difficult for environmental science, administration, and embryonic popular movements to develop.

After 1969 there was some opening for environmental agencies, following Soviet models. But throughout the Soviet system that trend was hampered by the fact that emissions controls were compelled to use obsolete Soviet technology, and contacts with western Europe's more advanced technology were still embryonic.

Tentative openings, 1972–1989

Major changes, officially launched by the 1972 United Nations Stockholm Conference on the Human Environment, began to evolve in the 1970s into environmentally stronger policy in many countries. Thereafter, international institutions under or related to the United Nations played gradually increasing roles in influencing environmental priorities in authoritarian as well as open governments, establishing non-politicized channels for international networks of scientists and planners across Cold War barriers.

In the Soviet republics this evolution was most evident in the German Democratic Republic (GDR), which developed working relations with international agencies and their West German counterparts well before the region-wide opening to the West in the late 1980s. The GDR established an Environment Ministry in 1972, just before its 1973 entry into the United Nations. By 1979 the international Convention on Trans-Boundary Air Pollution was a major step in the regime's liaison with western Europe. The GDR regime took greater advantage of the resulting opening to West Germany's advanced technology than Moscow allowed other East European countries to do. Public protests quickly followed, emerging rapidly in the mid-1980s; by 1988 there were widespread public protests. Much of that movement was under the protective umbrella of Christian churches, influencing the regime's tentative experiments in tolerance of public dissent. But censorship remained: emissions data were suppressed until late 1989.[2]

As the studies of Czechoslovakia and the GDR demonstrate, these contacts with western Europe became considerably more important than many environmental writers in the West knew before the collapse of the Soviet system.

Latin American dictatorships during the Cold War

Until 1990, access to international institutions and technologies was considerably more feasible for Latin American dictatorships than East European Soviet-structured governments. Through much of that era Guatemala, Brazil, Paraguay, and Chile were run by conservative military/civilian dictatorships supported economically and militarily by the United States. Brazil [1964–1985], Paraguay [1954–1989], and Chile [1973–1990] were all under military control. In Guatemala, this volume's example from Central America, a military coup displaced the civilian government in 1954, leading to open civil war from 1960 until 1996, under the heavy shadow of the Cold War and Washington's anti-Communist strategies. These countries underwent a variably complex evolution of the tension between extractive and environmental priorities, as they created a variety of environmental institutions, politically influential environmental science, and tolerance of public debate.

For purposes of comparing democratic with authoritarian regimes, it is important to recognize a central dynamic: each of these countries experienced continuity of development policy from the democratic governments of the early post-1945 years to the period of authoritarian rule and the return of democracy after about 1990.

One rationale of the dictatorships was that they were in a position to pursue development unhindered by the inefficiencies of open democracy. This volume's studies probe the consequences for the environment and the rise of countervailing socio-environmental movements as the years went on.

Brazil, the giant of South America, pursued a particularly massive effort under the generals to conquer the great Amazonia region and the southwestern interior, harnessing the vast hydro potential of its great rivers. This was a classic example of the claim of efficiency that dictatorships asserted as legitimization for massive industrial development oblivious to its environmental consequences.

Overcoming a long legacy of hostility, the military regimes of Brazil and Paraguay collaborated on building the massive Itaipu Dam on the Paraná River, a decade-long construction project from 1974 to 1984. Itaipu was a centerpiece of the energy policy that was critical for industrial development in a region poor in coal and oil resources. Government corporations, Eletrobrás in Brazil and ANDE in Paraguay, jointly managed the project, and construction was led by the Brazilian firm Mender Junior. Brazil's hydro-elite were dismissive of most environmental concerns, as most hydrological engineers throughout the world tended to be in the 1970s and 1980s.

Like massive hydro projects and river basin development projects around the world during the Cold War, the Itaipu Dam resulted in correspondingly great environmental degradation and displacement of indigenous populations. Only a combination of grass-roots resistance and international support could begin to mitigate the damage. Between them they developed rapid momentum after the dictators fell in 1984 and 1989. As in other Latin American countries, considerable protection against government repression was provided by Christian churches, the most powerful countervailing institution there: one wing of the Catholic Church in Latin America was under the drive of liberation theology, somewhat similar to the Protestant Church in GDR. Protest movements from the early 1970s, associated with international movements, in collaboration with a cosmopolitan scientific elite, began to influence the government's environmental standards, as expressed in the 1983 National Environmental Policy.

Within the binational project, planning and regulations developed from the mid-1970s onward under Eletrobrás management, in line with rising international pressure for including environmental concerns from the mid-1970s onward. The World Bank's policies were central: the Bank loaned more than US$3 billion to Brazil for its electricity sector between 1949 and 1987. Its environmental standards, strengthened greatly after 1983, placed increasing pressure on recipient governments.

In Chile, developmentalist technocracy dominated policy from the 1950s onward; throughout the following decades environmental distortions resulting from intensive industrial development continued. Serious air pollution conditions around Santiago and elsewhere prompted democratic governments to adopt environmental regulations in the 1960s, but efforts to enforce them had little success. For the copper industry, which dominated the country's exports, a government corporation built the Quintero Bay smelter on the central Pacific coast, supported by a highly polluting coal-fired power plant. After a coup in 1973, the new military regime led by General Pinochet, reflecting the copper industry's great political power, consistently suppressed evidence of air and water pollution, even as the smelter's socio-environmental costs became more and more evident. The regime nodded to environmental concerns in 1980 by adopting a constitutional right to a pollution-free environment, but failed to enforce it. Environmental opposition in Chile was expressed within the government, from without by international pressure, and locally by a public protest movement, but it gained minimal ground while Pinochet lasted in power.

Foreign investment and markets, primarily from the United States, were pervasive dimensions of the authoritarian regimes' power. Economists trained in the University of Chicago's neoliberal pro-market school, were influential in each regime. Though the "Chicago School" was only one school of economic development theory, it was (and still is) highly influential in politically conservative Latin American countries, emphasizing the development importance of attracting international corporate capital investment. This guiding assumption was also expressed through international banking institutions such as the United Nations Economic Commission for Latin America and the Caribbean, which had considerable leverage through its lending policies.

In these studies, international relations of the regimes and their societies were increasingly important for the environmental stakes from the 1970s onward. Throughout the Cold War, South American countries were more open to links with North American and western European democracies and capitalist corporations. Enormous infrastructure projects were managed by state corporations, with foreign investors from both the corporate sector and multinational governmental institutions.

Alliances with international organizations for environmental defense and social justice became highly important in the 1980s, as South American authoritarian regimes (more than their East European counterparts) became more sensitive to international pressure. This was dramatically true in Amazonia.

However, there could be another side to this collaboration: powerful non-government environmental organizations from the North often worked in uneasy collaboration with authoritarian regimes beginning in the late 1980s. Northern environmental institutes felt increasing urgency to protect tropical forest ecosystems. In Guatemala, the environmental aid establishment led by Conservation International worked with the regime's forest protection policy in the 1990s in creating the Mayan Biosphere Reserve.

Guatemala saw continuous development of the tropical forest region of Petén from the early 1950s onward. The military regime that seized power in 1960 used the pretext of systematically managing the forest to suppress potential political resistance by the Mayan minority there. The thirty-six-year civil war that followed is this volume's extreme example of turmoil and repression. But it is only an extreme instance of authoritarian forest management at the expense of local people that dates back to nineteenth-century Prussia in these studies.

Post-Cold War changes

The end of the Cold War saw transitions to regimes that were democratic in one degree or another in both East Europe and Latin America, but often with only slightly changing elites in command. There came to be greater space for civil society, open communications, dissent and debate, and public influence in determining governments' priorities—including environmental health, which local movements often experienced as public health crises. In these movements the social sources of resistance to authoritarian regimes included scientists, university students and faculty, farmers, and a range of non-governmental organizations.

In the 1990s, Latin America saw transitions from military-supported dictatorships to variably more democratic regimes, but with many continuities of control by economic and political elites.[3] Where economically conservative governments came to power, they maintained neoliberalism's alliances between economic elites and international capital; market-based mechanisms for managing environmental issues limited the roles of law and public administration.

In eastern Europe, as the Soviet system collapsed, the sudden efflorescence of grass-roots protest movements expressed nationalism in environmental guise, with short-lived citizens' environmental activism.[4] The successor regimes trended away from state socialism to neoliberal nationalist capitalism. However, over the following quarter century, there were authoritarian continuities among ruling circles, which have emerged more clearly in recent years. In the ideological vacuum left by the collapse of the Soviet system, authoritarian elites revived ultra-conservative nationalism in Poland, Hungary, and elsewhere. In the Białowieża Forest region of eastern Poland, particular tree species chosen for reforestation to replace massive logging have become the regime's symbols of strident Polish nationalism; the combination is undermining sustainable forest ecology.

Colonial legacies: a note on Africa and South and Southeast Asia

This volume has far less coverage of the other great trend of the post-1945 decades: the varied transition from West European colonial empires to subsequent independent governments in many countries of sub-Saharan Africa and South and Southeast Asia. The legacy of European colonial regimes was weighty not only for the evolving spectrum of democratic and authoritarian regimes, but also for continuity and change in administrative agencies such as forestry, where policies pertaining to environmental impacts were lodged.

Those young governments varied widely from one country and time period to another. This collection is represented by Zimbabwe in the years since 1980, when a newly independent government replaced colonial Southern Rhodesia. The achievement of independence there gave great hope—as it did in adjacent South Africa in 1994—for establishment of racial and social equity, with its potential for environmentally sustainable development. But this hope rested upon the political stability of Robert Mugabe's regime, which became increasingly challenged after 2000. From then on it devolved into a new authoritarianism, with great personal wealth for the inner circle and severe ecological costs to the nation.

The bottom line

These studies show a wide range of complexity among authoritarian regimes—both state-socialist and right-wing dictatorships. Until the 1970s the environmental dimension was weak in every one of these countries.

Ecological science and its research institutions were still becoming mature. Environmental priorities were still embryonic in institutional settings, in no position to counter-balance the dominant developmentalist priority. When that began to change, the challenge for each regime, especially the authoritarian countries, was how and whether to expand the scope of environmentally responsive policy, and how and whether to open to public participation the setting and implementing of environmental values.

The reverse recent trend toward more authoritarian governance in eastern Europe and elsewhere makes this collection highly relevant for our understanding of the environmental stakes in today's world.[5] It is a sobering trend, as each chapter has discussed the limits of environmental concerns in authoritarian regimes, and the importance of three dimensions: the strength of environmental agencies against their adversaries within regimes, the public support of environmental priorities in the political process, and the importance of international environmental institutions and NGOs.

Overall, though considerably more in authoritarian systems than in open societies, there has been great difficulty developing environmental law and agencies, non-governmental institutions and environmental organizations, or sustaining citizens' activism.

Notes

1 See, for example, Stephen Brain, "Stalin's Environmentalism," *The Russian Review*, 69, no. 1 (2010): 93–118.
2 For Hungary in the same era, see Viktor Pál, *Technology and the Environment in State-Socialist Hungary: An Economic History* (New York: Springer, 2017).
3 David Pion-Berlin and Rafael Martinez, *Soldiers, Politicians, and Civilians: Reforming Civil-Military Relations in Democratic Latin America* (Cambridge: Cambridge University Press, 2017).
4 Jane I. Dawson, *Eco-nationalism: Anti-nuclear Activism and National Identity in Russia, Lithuania, and Ukraine* (Durham, NC: Duke University Press, 1996).
5 For a pessimistic view for Southeast Asia, with implications for other regions, see Mark Beeson, "The Coming of Environmental Authoritarianism," *Environmental Politics* 19, no. 2 (2010): 276–294.

Index

Page numbers in **bold** denote tables, those in *italics* denote figures.

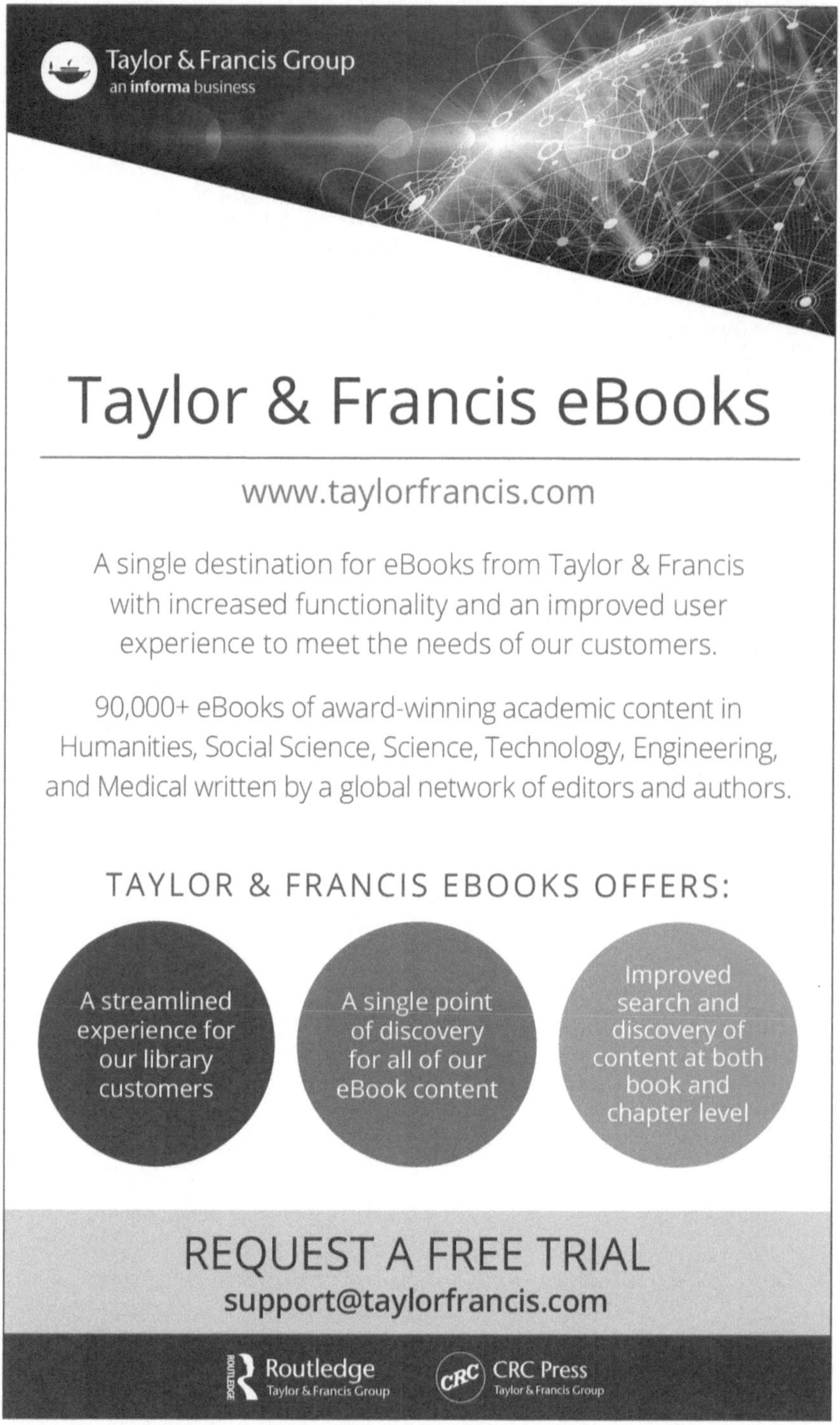
Taylor & Francis Group
an informa business
Taylor & Francis eBooks
www.taylorfrancis.com
A single destination for eBooks from Taylor & Francis with increased functionality and an improved user experience to meet the needs of our customers.
90,000+ eBooks of award-winning academic content in Humanities, Social Science, Science, Technology, Engineering, and Medical written by a global network of editors and authors.
TAYLOR & FRANCIS EBOOKS OFFERS:
A streamlined experience for our library customers
A single point of discovery for all of our eBook content
Improved search and discovery of content at both book and chapter level
REQUEST A FREE TRIAL
support@taylorfrancis.com
Routledge
Taylor & Francis Group
CRC Press
Taylor & Francis Group